Polymers in Conservation

Polymers in Conservation

Edited by

N.S. Allen, M. Edge
Centre for Archival Polymeric Materials, Manchester Polytechnic

and

C.V. Horie
Manchester Museum, University of Manchester

ROYAL
SOCIETY OF
CHEMISTRY
Information
Services

The Proceedings of an International Conference organized by Manchester Polytechn
and Manchester Museum, Manchester, 17th–19th July 1991

Special Publication No. 105

ISBN 0-85186-247-0

A catalogue record for this book is available from the British Library

Published by The Royal Society of Chemistry,
Thomas Graham House, Science Park, Cambridge
CB4 4WF

Printed in Great Britain by Bookcraft (Bath) Ltd.

Preface

A major triumph of the industrial revolution was the introduction of new materials. Building on the wealth of natural products from the living world, innovators extracted and modified these starting materials to create usable products. Organic chemistry progressed, enabling materials to be constructed synthetically. The process of innovation, application, and occasionally redundancy continues apace.

Many of these materials are polymers, such as fibres, plastics, adhesives and coatings. Others are supplementary to the polymers, such as dye and plasticiser additives. Incorporated in objects, these form the material evidence for much of human history and discovery over the past two centuries. Preservation of the objects depends on the survival of the materials of which they are composed.

All organic materials are subject to deterioration from light, oxygen and water and would be expected to have a limited life even if kept in good conditions. After decades of use and poor storage, the polymers deteriorate and the objects collapse, sometimes catastrophically. There is an increasing rate of loss of cinema film, early plastic utilitarian objects that changed our way of life, and innovative sculptures in plastic. Recently plastics have been designed to self-destruct or for recycling, with severe consequences for their long term survival as evidence.

Until recently objects have been examined and considered by their morphology and appearance. Increasingly, information is being sought from objects at a molecular level, for example in authenticity or technology studies. The need is most striking in natural history where biopolymers, particularly DNA, are extracted and used as primary evidence.

Improved techniques of storage are required in order to reduce changes in the polymers. Processes of deterioration must be understood to specify these techniques and also to reconstruct the original state from the altered survivals.

The Centre for Archival Polymeric Materials has carried out innovative research in this field and has welcomed the increasing recognition of the problems world wide. This volume brings together for the first time workers in all these aspects of a common problem. The authors also represent the life cycle of an object, from the academic scientist understanding and developing new materials, through the manufacturer, to the private collector and finally the museum. Only by such collaboration will the achievements of recent history be preserved for posterity.

C.V.Horie
The Manchester Museum

Acknowledgements

The Congress from which these papers were developed benefited greatly from a broad range of sponsors: Fuji Photofilm Co., Kodak (UK) Co., The National Film Archive, ITN News, Ciba-Geigy Corp, Adeka-Argus Chemical Co., Asahi Chemical Ind. Co., The Conservation Unit of the Museums and Galleries Commission, and the Greater Manchester Museum of Science and Industry.

The members of the Centre for Archival Polymeric Materials wish to thank colleagues who assisted with the planning and organisation of the Congress: G.Boston (BBC), K.Brems (Agfa-Gaevert), D.Fromageot (Laboratoire de Recherche des Musées de France), M.Gomez (North West Film Archives), H.Iwano (Fuji Photofilm), A.Moncrieff (Science Museum), D.Pullen (Tate Gallery), T.Ram (Eastman Kodak), and C.Williamson (Plastics Historical Society).

The Centre for Archival Polymeric Materials
John Dalton Building
Manchester Polytechnic
Chester Street
Manchester M1 5GD, UK

C.V.Horie (The Manchester Museum, University of Manchester), Dr.M.Edge and Prof.N.S.Allen (Department of Chemistry, Manchester Polytechnic), and Prof.T.S.Jewitt (Graphics Technology, Manchester Polytechnic).

Contents

150 Years of Plastics Degradation

C. J. Williamson

PLASTICS HISTORICAL SOCIETY, C/O THE MANSION HOUSE, FORD, SHREWSBURY, SHROPSHIRE SY5 9LZ, UK

"If nature hadn't produced plastics, then man would have had to invent them."

So great is the influence that plastics have on modern living that life without them would be impossible, with approximately 100 million tonnes produced each year and most human activities relying on them to a lesser or greater extent. From electrical insulation, enabling telecommunications, computers, electric motors and power in the home and industry, and packaging saving millions of tons of foodstuffs from decomposing before reaching the consumer, to artificial heartvalves and heat insulation in homes, plastics enable western society to live and to grow at the rate to which we have become accustomed.

It was man's insatiable drive for growth which led the desire to tame, modify and control nature in a search for new worlds, power and above all, profit. By the mid 19th century the profit motive had driven man into the industrial revolution by the application of engineering and rudimentary science. Many of the original village craft skills had been transferred to large manufacturing industries producing massive volumes at low costs. The village spinner and weaver had lost his business to the textile mills in Lancashire and Yorkshire. The blacksmith's customers now bought goods from the 'metal bashing' Black Country leaving him only as a shoer of horses, and for decades the potter had given way to the Wedgwoods of Stoke-on-Trent. Fortunes had been made by the new industrialists and anything which could be mechanised was seen as yet another potential fortune to be made.

Wood turning and carving with wood, bone and horn to produce utensils and small decorative items were still village crafts, but when the new inventors and industrialists turned their attention to this market then its decline was inevitable. Man had learned how to mechanise the moulding of ceramics and metals, but wood, though carvable, was almost impossible to mould. However, some natural carvable materials were also mouldable and indeed had been moulded for centuries.

The earliest civilisations had used animal horns as liquid containers and it is inconceivable that they had not also discovered that boiling water or heat from a fire made the horn soft and pliable. By 1710 Obrisset in London was making snuff boxes by moulding horn using heat to soften the 'thermoplastic' material.(1) He clamped the hot horn (or hoof) in a prewarmed steel mould, and allowed them to cool to room temperature. On splitting the male and female mould halves the resultant 'moulding' was released and required only edge trimming before being mated to a similar snuff box half. Thus an apparently carved snuff box lid could be produced in say 20 minutes, a top which fitted exactly to the moulded base every time, and the resultant saving in time compared to his contemporaries should have made him a wealthy man. Even before Obrisset, sheets of thin horn were used as wind shields in early lanterns - originally known as 'lant-horns'. (2)

Most early horn mouldings were produced in shallow relief, but with the popularisation of jet in the 1860's following Albert's death, hoof and horn were widely used for producing brooches, buckles, bracelets and many other small items of decoration. Horn is relatively stable under 'normal' storage conditions but many brooches can be found exhibiting characteristic splitting as the moulded-in stress is partially relieved, especially if the article has been exposed to heat. (Figure 1)

Other natural polymeric materials can be moulded but many require some degree of chemical or physical modification before they can become practically usable. Shellac, an exudate from the lac insect, is thermoplastic but brittle and formed the basis of many mouldable compositions in the 19th century. Most frequently found as Union cases, a stable material was produced in the 1850's and 60's in the USA, compounded with wood flour. Union cases protect daguerreotypes and ambrotypes from light and the shallow relief mouldings exhibit good stability. Other shellac based compositions (eg Manton ware, Lionite etc) also appear to be stable although most are prone to physical damage such as chipping and cracking. Shellac crosslinks on moulding or heating and after prolonged storage. (3)

Bois Durci, patented in France in 1855 is a mixture of powdered wood and albumin from blood, moulded under pressure with steam heating.(4) Excellent quality in both relief and three dimensional moulding were made and these products exhibit good stability (Figure 2).

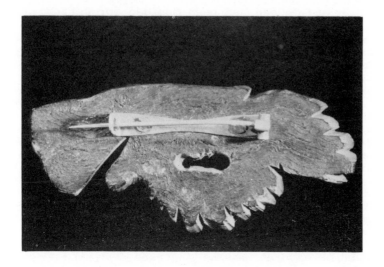

Figure 1. Reverse of horn/hoof brooch showing characteristic splitting

Figure 2. Bois Durci photograph frame.

Natural rubber is certainly a mouldable material but becomes more useful when vulcanised with sulphur. With increasing levels of sulphur, vulcanised rubber exhibits increasing rigidity and hardness up to about 30% when vulcanite (ebonite or hard rubber) is produced. Vulcanite was also used from the 1860's to produce dark coloured brooches and small decorative items as well as an electrical insulator. Black was the simplest and most effective colour as a good gloss and jetness could be made, but is vulnerable to surface degradation on exposure to moisture and light, producing (indirectly) a sulphuric acid rich dull grey-brown surface layer (Figure 3, 4). This of course will attack other materials in close proximity although vulcanite itself is relatively stable to mineral acids and the degradation does not appear to be progressive. Rubber itself is prone to oxidation and is especially sensitive to ozone and catalysed by the presence of metals (eg copper). (5)

Gutta percha, the trans isomer of rubber, is also a mouldable natural polymer extracted from trees with the advantage over rubber that it is inelastic and more rigid at room temperature, softening with boiling water for easy moulding. The basic technology of plastics extrusion was developed around gutta percha, including wire coating for undersea telegraphic cables. (6) Gutta

Figure 3. Vulcanite bracelets, upper in original condition, lower showing characteristic greyish (green-brown) surface following exposure to moisture and light.

Figure 4. Vulcanite pin box showing original
condition inside.

percha embrittles with time (Figure 5) and even modern
thin sheet material can degrade in less than 2 years
under normal storage conditions.

Figure 5. Gutta Percha 'medal' showing embrittled,
friable surface.

So great was the interest in producing a mouldable alternative to carving that in 1858 approximately 8 % of all British patents mentioned the words 'mouldable composition', or 'plastic'. Many were based on shellac or gutta percha and necessarily overlapped, but others were based on more unusual natural materials e.g. peat, seaweed and powdered leather.

The earliest semi-synthetic plastics, chemically modified natural polymers, are essentially unstable especially under unventilated conditions. Photographic film is the most researched and widely reported, but solid products are also vulnerable despite a slightly lower level of nitration. (7,8) Degradation is first observed as an acidic surface bloom (Figure 6) but by this stage the material is already seriously degraded. Some mouldings are more susceptible than others and with feedstock variations in the original polymer (natural cotton) and unrecorded production and use conditions it is difficult to establish the reasons for the variation. However, transparent celluloid (cellulose nitrate) appears to be particularly unstable, probably because early opaque mouldings commonly included zinc oxide (a white pigment), acting as an internal stabiliser. (9) The build-up of acidic degradation products is critical as they promote further degradation and their removal is essential to ensure longevity. Thick sections tend to degrade before thin ones (Figure 7) and the presence of metals speeds the reaction. Light is also reported to promote the degradation.

Figure 6. The first signs -- strongly acidic
          'bloom' on the surface of degrading
          cellulose nitrate.

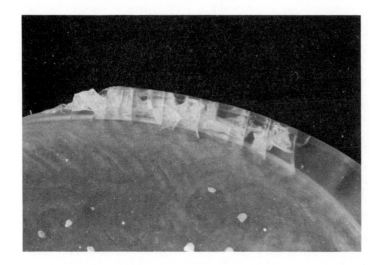

Figure 7. Cellulose nitrate 'crazing', initially
         in thick sections.

    Following the surface bloom, the mouldings start
to craze and frequently discolour and eventually
complete disintegration occurs (Figure 8).

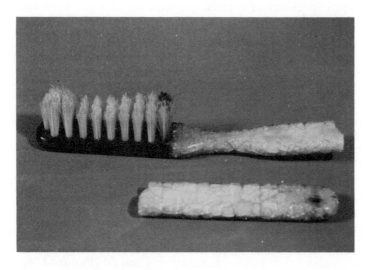

Figure 8. Disintegrating cellulose nitrate
         tooth-brush.

Figure 9. Cellulose acetate toy, distortion due to
        plasticiser loss.

        The development of 'non-flam celluloid' or
cellulose acetate produced a more stable plastics
material and plasticiser loss is the most frequently
observed problem. This results in distortion (Figure
9), embrittlement and eventual complete degradation.
(10) Again the stability of cellulose acetate filmstock
is well reported and the acidic degradation products
will both promote further degradation and attack other
nearby materials.

        Storage conditions of cellulose esters are clearly
critical to their longevity and it is suggested that a
possible (although probably impracticable) answer might
be a dry deep-freeze to slow down the reaction. Under
more normal conditions acidic degradation products
must be allowed to escape, but must not be allowed to
affect nearby similar materials. Clearly this is
difficult to realise in practice and collections must
be regularly examined for early signs of degradation
and any affected article isolated.

        It is possible that the use of an acid absorbing
coating might delay degradation by 'mopping-up' acidic
products, and initial trials have been conducted with
encouraging results using epoxidised soya bean oil
(ESBO). A mirror back, probably from the 1920's
exhibited the clear signs of degradation in the thick
rim section in four distinct areas (Figure 7). The
object was half immersed in ESBO and left at room
temperature for approximately 2 years. The increase in

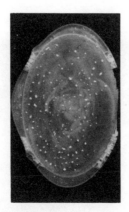

Figure 10. Before (left) and after (right) ESBO
treatment to lower half for 2 years.
Note the increased crazed area to
top left compared with bottom left.
(Degraded area to right removed for
analysis)

the crazed area was taken to be an indication of
continuing degradation. That immersed in the ESBO had
not changed whereas the other areas had approximately
doubled in area and the intensity of crazing had
increased (Figure 10). Other objects already exhibiting
the acidic surface bloom have been smeared with ESBO
and the initial results indicate a slowing down of the
degradation. Whilst this technique is unsuitable for
objects on display it may be applicable to those in
storage, especially as a precautionary measure where
regular inspection cannot be guaranteed. The 60% plus
loss of the Vestry House Museum collection of Halex
products, (11) and the ongoing loss of specific objects
in other museums (Science Museum, Oyonnax etc) and
collections should be sufficient incentive to promote
further investigative work with this system.

Casein, the other semi-synthetic plastic and
commonly found as buttons, uses water as a plasticiser
and under normal conditions of use can alternatively
absorb and release moisture from its environment,
swelling and contracting and eventually forming surface
crazing typical of casein mouldings (Figure 11).
Constant humidity levels in prolonged storage are
therefore important. (12)

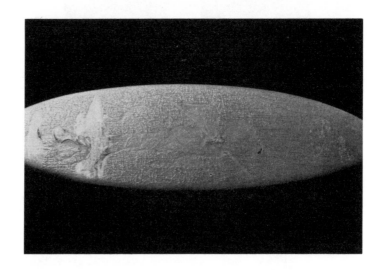

Figure 11. Casein brush handle showing
characteristic surface crazing.

    With the introduction of fully synthetic polymers
in the early part of the 20th century a new generation
of plastics was born with a new range of stability
problems. Baekeland controlled the reaction between
phenol and formaldehyde to enable Bakelite lacquers,
casting and moulding materials to be produced. Phenolic
can be cast as a water-white moulding but early
examples yellowed and darkened on exposure to light,
and some Damard Lacquer Company examples are now
completely black although retaining their surface
gloss. Shades of amber are the most frequently found
although some of these may have originally been
produced untinted. Similarly, blue examples frequently
have a green surface due to yellowing.

    In moulded form, mixed with wood flour, prolonged
exposure to light causes a loss of surface gloss and
occasionally a fading of the pigments (Figure 12). The
lighter coloured and later urea formaldehyde shows few
problems of stability, but thiourea/urea used in picnic
sets frequently becomes surface stained and crazed
following use with hot water (especially tea cups !).

    Post war, oil derived polymers are not free from
problems and early polyethylene bottles of the
'squeezy' type are starting to increase in rigidity.
Polyethylene film is sensitive to light with one year
outdoor exposure a typical limit. Polypropylene is
equally affected, occasionally accelerated by

Figure 12. Phenolic ash tray set showing loss of
          gloss and pigment fading on prolonged
          exposure.

inappropriate pigments and early beer bottle crates are
frequently found with friable surfaces. Developments in
stabilisation of polyolefins have allowed far more
light stable mouldings to be made with an outdoor life
expectancy of several years.

        Some extreme use conditions may also trigger
degradation, for example medical equipment sterilising
trays in polyethylene quickly became brittle due to
high energy radiation levels used in sterilisation.

        The recent concern for the environment has also
stimulated the production of 'biodegradable' plastics
which although environmentally unsound have had a
limited market success and must present a new range of
stability problems to the conservator and curator.

        PVC was for many years unprocessable as a rigid
plastics material as it decomposes at approximately 180
C, compared to a softening point of about 160 C. The
development of copolymers, stabilisers and lubricants
allowed the rigid PVC industry to start. Examples of
under stabilised corrugated roofing sheet occur,
transformed from a clear, tough transparent sheet to
brown, opaque brittle and broken pieces (Figure 13).

        Other polymers have varying stability, especially
to u.v. light, solvents, stress and heat. PMMA
(acrylic) is frequently used in three dimensional art

Figure 13. Inadequately stabilised rigid PVC
roofing, brittle and opaque after
exposure.

forms, jewellery, advertising and lighting, but if not
annealed will surface craze after relatively limited
outdoor exposure.

Frequently products and plastics raw materials
were imported into this country in the 1950's and
1960's from low technology supply sources and normally
these are indistinguishable from well produced items.
These are probably more vulnerable to the normal
degradation initiators and regular inspection of
collections is recommended.

REFERENCES

1. P.A.S. Phillips, 'John Obrisset', Batsford, 1931
2. P.Hardwick, 'Discovering Horn', Lutterworth,
    Guildford, UK, 1981
3. B.S.Gidvani, 'Natural Resins', IPI, London, 1946
4. Brit. Pat. 2232, 1855, F.C.Lepage
5. J.R.Scott, 'Ebonite', Maclaren , London, 1958
6. G.L.Lawford & L.R.Nicholson, 'The Telcon Story',
    Telcon, London, 1950
7. C.Selwitz, 'Cellulose Nitrate in Conservation',
    Getty Conservation Institute, California, USA, 1988
8. N.S.Allen et al, J. Photogr. Sci., 1988, 36(2), 34
9. E.C.Worden, 'Nitrocellulose Industry', Constable,
    London, 1911

10. C.R.Fordyce & L.W.A.Meyer, <u>Industr. Engng. Chem.</u>,1940, <u>32</u>, 1953
11. A.Wright, <u>plastiquarian</u>, 1989, <u>2</u>, 6
12. E.Sutermeister & F.L.Browne, 'Casein', Reinhold, New York, 1939

Fundamental Aspects of Polymer Degradation

Ian C. McNeill

POLYMER RESEARCH GROUP, CHEMISTRY DEPARTMENT, UNIVERSITY OF GLASGOW,
GLASGOW G12 8QQ, UK

INTRODUCTION

Effects of Polymer Degradation

The occurrence of polymer degradation is recognised by its effects on the appearance and properties of the material. Common effects are discoloration, embrittlement, tackiness, loss of surface gloss or crazing or chalking of the surface. Under more extreme conditions, release of volatile products or burning may be observed. From a practical point of view, degradation is any change which has an adverse effect on polymer properties. To the polymer chemist, the term "degradation" refers to any chemical change in the polymer structure.

Although degradation is usually regarded as an undesirable effect, it is worth noting that there have recently been developed some very positive applications of polymer degradation, in controlled lifetime polymers (particularly for agricultural applications) and in microlithography for the electronics industry.

Nature and Types of Macromolecule

The chemical changes involved in polymer degradation have to be considered in relation to the structure of the polymer and it is useful at this stage to survey briefly the various types of macromolecular structure present in familiar polymers. It is possible to classify polymers in various ways, based, for example, on their method of preparation (addition or step-growth polymerisation) or their physical properties (elastomers, resins and fibres), but for the purposes of this discussion I would like to stress the following possible divisions.

Firstly, one could divide polymers into *natural polymers* of plant, animal or insect origin and *synthetic polymers*. A second important distinction, which proves to be important in relation to degradation behaviour, is based on the repeat structure. Macromolecules are made up of many chain units which are repeated many times along the length of the chain. Most polymers contain a single type of repeat unit and are described as *homopolymers*. There is, however, a large group of polymeric materials (both natural and synthetic) in which two or more different structures are repeated in the chain: these are described as *copolymers* and include such examples as some synthetic rubbers and lacquers, and natural protein-based materials. Also very important in relation to degradation behaviour is the distinction between polymers with a purely carbon-carbon backbone and those with heteroatoms.

The structures of some important polymers are grouped for illustration in Figures 1-3. Taking the **carbon chain** and **heterochain** distinction as the initial basis of classification, Fig. 1 shows examples of carbon chain polymers, including only one natural polymer, *cis*-polyisoprene (natural rubber). All the materials illustrated are homopolymers, except for the styrene-butadiene copolymer. In this, the two types of repeat unit appear at random along the chain, although the overall proportions may be controlled. With high butadiene content, the product is a synthetic rubber, whereas with high styrene content an impact resistant resin is obtained.

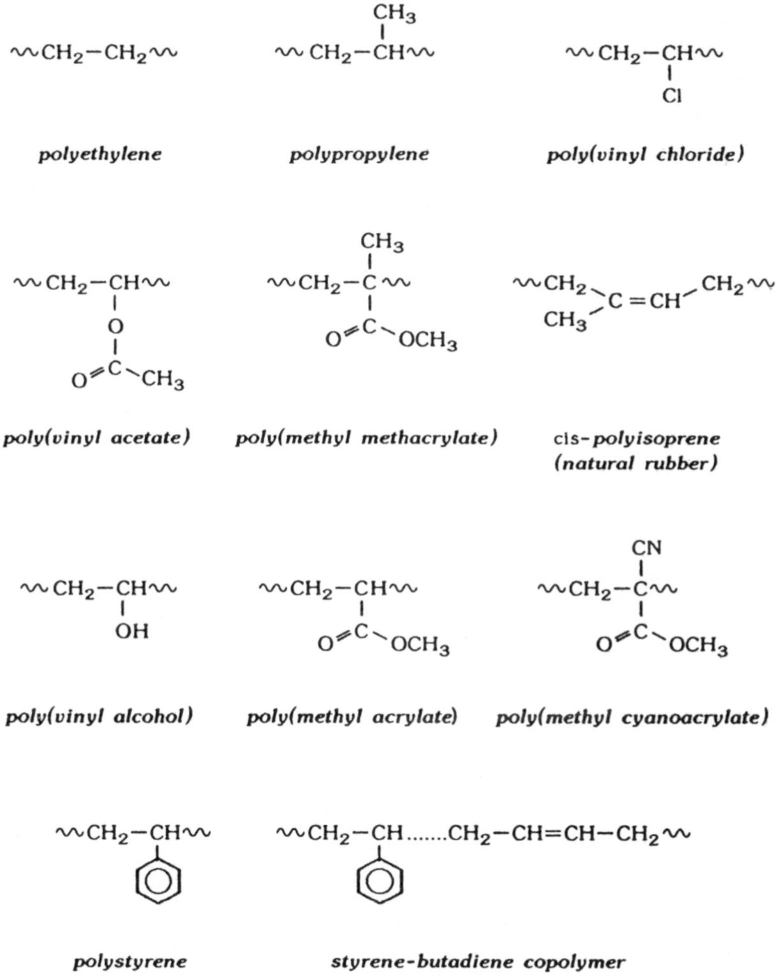

Fig. 1. **Some Important Carbon Chain Polymers**

Heterochain polymers have more complex structures. Figure 2 illustrates examples of natural polymers. Those of animal origin are usually proteins and have a polypeptide chain structure. This is a special type of copolymer structure which is non-random and based on various repeat units, derived from α-aminoacids with different R groups. Wool, silk, casein and albumin are examples of materials with this type of structure. Those of plant origin will typically be polysaccharides with structures related to the structures of cellulose or starch illustrated. Cellulose acetate and cellulose nitrate are derived from cellulose by replacement of hydroxyl groups by acetate or nitrate groups, respectively. Shellac is an interesting example of a polymeric material of insect origin: it has a very complex structure which is not fully understood, but includes chain structures of the type shown. All of these materials have the common feature that the backbone of the macromolecule includes other atoms as well as carbon, here oxygen and nitrogen.

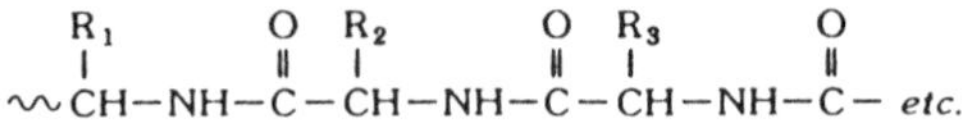

$$\sim\!\!\sim\!\!\underset{\underset{\displaystyle R_1}{\displaystyle |}}{CH}\!-\!NH\!-\!\underset{\underset{}{\displaystyle \|}}{C}\!-\!\underset{\underset{\displaystyle R_2}{\displaystyle |}}{CH}\!-\!NH\!-\!\underset{\underset{}{\displaystyle \|}}{C}\!-\!\underset{\underset{\displaystyle R_3}{\displaystyle |}}{CH}\!-\!NH\!-\!\underset{\underset{}{\displaystyle \|}}{C}\!-\ etc.$$

polypeptide chain structure in a protein

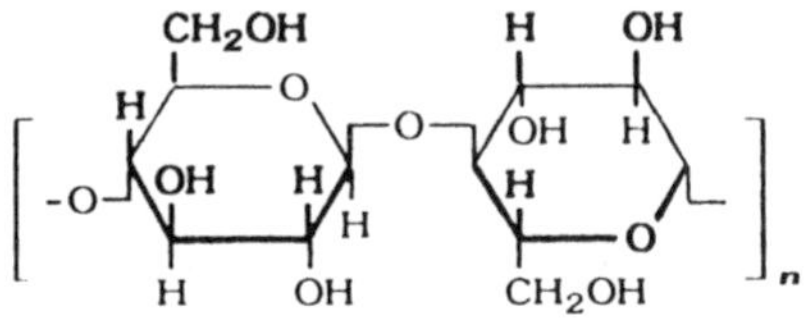

cellulose 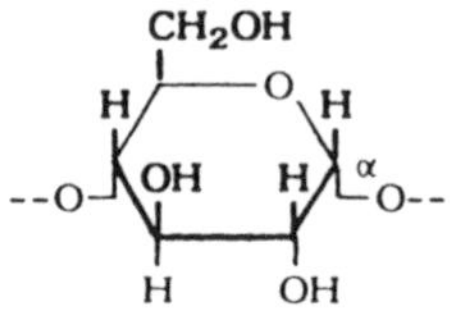   *starch*

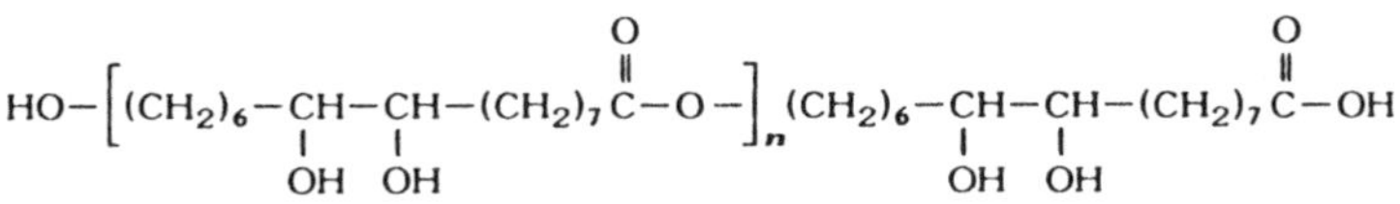

$$HO-\left[(CH_2)_6-\underset{\underset{\displaystyle OH}{|}}{CH}-\underset{\underset{\displaystyle OH}{|}}{CH}-(CH_2)_7\overset{\overset{\displaystyle O}{\|}}{C}-O-\right]_n(CH_2)_6-\underset{\underset{\displaystyle OH}{|}}{CH}-\underset{\underset{\displaystyle OH}{|}}{CH}-(CH_2)_7\overset{\overset{\displaystyle O}{\|}}{C}-OH$$

part of the structure present in shellac
(other long chain and terpinic acid residues may be present)

Fig. 2. Some Important Heterochain Polymers: a. Natural Polymers

Synthetic heterochain polymers (Fig. 3) range from relatively simple structures such as that of the polyamide, Nylon-6, and the silicone rubber, polydimethylsiloxane, to very complex structures such as are present in alkyd and epoxy resins. All the polymers previously mentioned have a structure of linear chains, but the alkyds and epoxy resins differ in being three-dimensional networks. Some of these materials also have the feature of aromatic rings built into the polymer backbone.

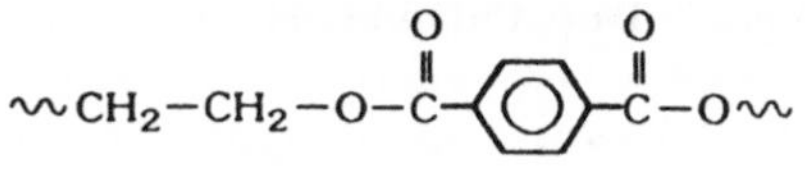

poly(ethylene terephthalate)

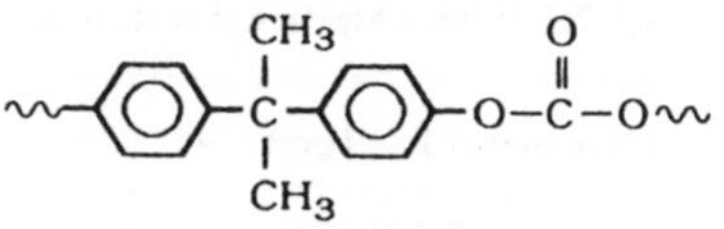

a polycarbonate

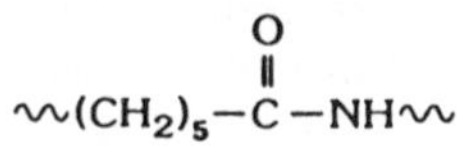

a polyamide (Nylon-6)

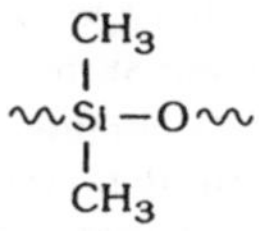

polydimethylsiloxane
(a silicone rubber)

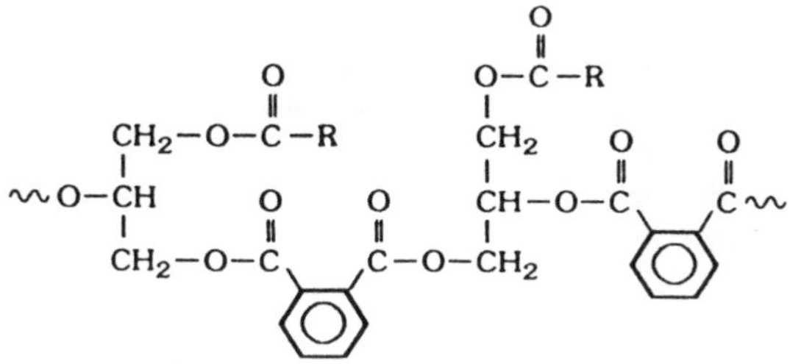

a typical
alkyd resin

(R contains
unsaturation)

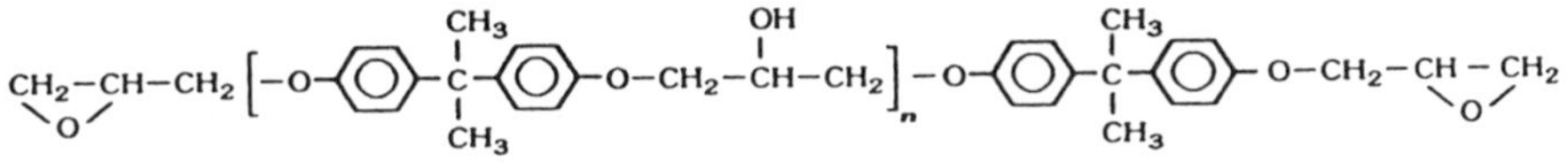

linear epoxy resin prepolymer (oligomer)

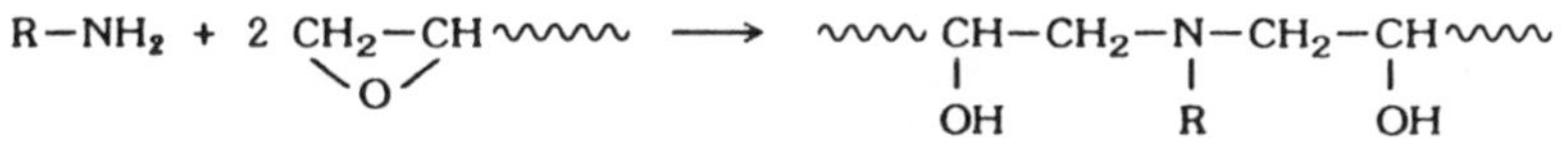

amine cured epoxy resin

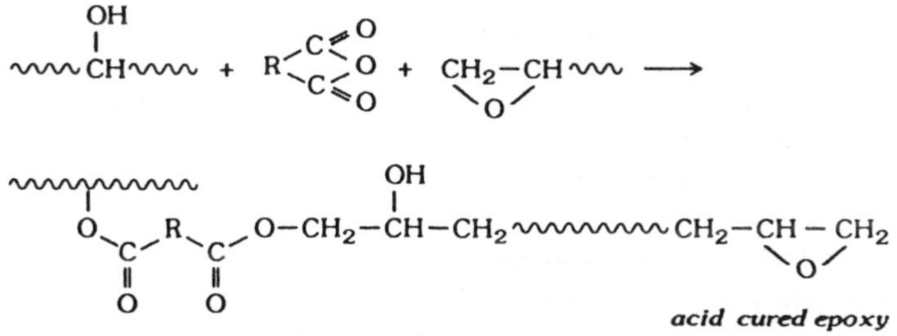

acid cured epoxy resin

Fig. 3. Some Important Heterochain Polymers: b. Synthetic Polymers

GENERAL ASPECTS OF POLYMER DEGRADATION

Degradative Agencies

Polymers may be degraded as a result of exposure to various agencies, which can be listed under five headings:

1. *Heat.* Every polymer will undergo degradation at some stage if the temperature is increased sufficiently and the differences in stability are considerable. For example, PVC begins to discolour due to thermal degradation below 200°C, whereas polytetrafluoroethylene is stable to near 500°C. The relative thermal stability of some widely used polymers is illustrated in Fig. 4, by comparing the temperature needed to give 15% weight loss, in the absence of air, when the temperature is gradually raised. It has to be stressed, however, in relation to comparisons such as this, that polymer stability can be dependent on the history of the sample, its purity and, in some cases, on molecular weight. As well as being important in relation to processing and use, thermal degradation is very relevant to the behaviour of polymeric materials in the extreme situation of a fire.

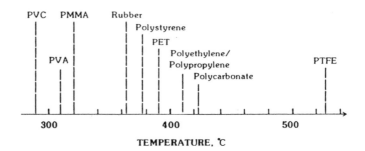

Fig. 4. **Relative Thermal Stability of Some Common Polymers: Temperature for 15% Weight Loss in Nitrogen at 10°C/min.**

2. *Light.* The spectrum of radiant energy from the sun which reaches the surface of the earth in summer extends from about 290 nm into the infrared region and the ultraviolet component of this radiation is sufficiently energetic to break bonds and cause degradation in many polymers, as will be considered subsequently. Polymers are even more sensitive to more energetic radiation such as X-rays and γ-rays.

3. *Atmosphere.* Degradation of polymers can also occur by chemical attack by gases in the atmosphere. The most important case is attack by oxygen, but aggressive gases such as ozone, sulphur dioxide and nitrogen dioxide can have an important effect even in low concentrations. Oxidation often accompanies the primary effects of degradation induced by heat and light, especially in the surface layer of the polymer, in the common situation of degradation in air.

4. *Hydrolysis.* Certain polymers are susceptible to this specific type of chemical attack under appropriate conditions.

5. *Biodegradation.* Most natural polymers, but relatively few synthetic polymers, are sensitive to attack by fungi and microorganisms, again under appropriate conditions.

Some Important Polymer Characteristics

Some general features of particular importance in relation to degradation processes may be highlighted at this point.

a. *Amorphous or crystalline state.* Most polymers are amorphous solids, i.e. they have a very disordered structure in which the chains are randomly coiled. It is easier for atmospheric gases to diffuse into the surface layer of an amorphous polymer, and for gaseous products of degradation to escape, than in the case of a crystalline polymer.

b. *Glass transition temperature.* This is a very important property of any amorphous, linear polymer. If the temperature is gradually raised, this is the temperature at which the physical state changes from glass to rubber. The change is reversible on cooling. At the glass transition temperature (T_g), the macromolecules receive just enough thermal energy from the surroundings for movement to begin within the chains. Some types of degradation reaction cannot proceed to any significant extent below T_g, because the macromolecules are "frozen" in their random coils.

c. *Functional Groups Present.* Functional groups are specific chemical structures within the polymer chain which provide sites for reaction. In the various structures shown in Figs. 1-3, examples of functional groups are carbon-carbon double bonds, -OH groups and -CO-O- (ester), -CO-NH- (amide) and -O-CO-O- (carbonate) linkages. Polymers containing ester and amide linkages may be susceptible to hydrolysis, for example, and those with OH substituents may undergo dehydration at elevated temperatures. Structures containing the carbonyl (-CO-) group are also particularly important as *chromophores,* i.e. sites at which certain types of light may be absorbed.

d. *Presence of Tertiary Hydrogen Atoms.* The CH groups in polypropylene, polystyrene, polyacrylates and various other polymers are sensitive to certain types of reaction during degradation, in which free radical species are present.

Stability and Degradation Behaviour

It is possible to predict sites of chemical attack in reactions such as hydrolysis in suitable polymers, from examination of available functional groups, but it is much more difficult to predict general patterns of degradation due to heat and light. It might be considered that a good starting point would be an examination of bond energies as a guide to which bonds would be first to break. Some typical average values of bond energies are shown in Table 1.

Table 1. **Average Bond Energies at 298K** *(in kJ mol⁻¹)*

C—S	273	Si—Si	226	P—N	200	C=C	612
C—Br	280	Si—H	319	P—C	264	C=N	617
C—N	307	Si—O	432	P—H	322	C=O	732
C—Si	328	N—H	391	P—O	360		
C—Cl	340	O—H	464				
C—C	349						
C—O	361						
C—H	416						
C—F	485						

The unreliability of this approach can be illustrated by just two examples. It would be expected from bond energy considerations that silicone rubber should initially undergo degradation by scission of C—Si bonds, whereas what is observed is that backbone Si—O bonds break. Since the C—Cl and C—C bond energies do not differ greatly, there would be no reason to expect PVC, on this basis, to be much less stable than polystyrene or polyethylene, and the stability of low molecular weight alkyl halides supports this view. Yet the stability comparison in Fig. 4 shows PVC to be 100 or more degrees less stable than these two polymers.

Clearly there are other important factors involved in polymer degradation, apart from bond energies. Several such factors need to be considered. Firstly, there is the fact that a macromolecular chain is a unique reaction environment. In most chemical reactions, in order for reaction to occur, reactant molecules must collide and do so with sufficient energy. In the long chain of a macromolecule, however, repeat units are held in proximity to one another. This has two consequences. Suitable functional groups in adjacent repeat units may be able to react. It may also be possible for a reaction to proceed systematically *along* the chain. The physical state of the polymer is also relevant to degradation behaviour. Degradation induced by light can take place in the solid state and that brought about by heat will occur either in the solid state or in a viscous melt. Such reaction conditions are very different to those encountered in many reactions between small molecules.

Secondly, the stability of a long macromolecule can be acutely sensitive to the presence of a very small number of structural abnormalities, where deviation from the repeat structure occurs. The simplest such abnormality is the chain end itself and there are important examples of end-initiated thermal degradation. Other common abnormalities are oxidised structures, possibly introduced during processing, and these can be sites both for thermal and photolytically induced degradation.

Thirdly, polymers are seldom used in the pure state and impurities or additives can have a profound effect on the stability of a polymer and on the pattern of degradation. Additives such as stabilisers and fire retardants have a very specific function in relation to degradative processes, but other additives such as fillers, dyes and colourants may also have an important role.

Basic Reaction Types

Degradation reactions in polymers fall into two broad categories. Since they tend to take place either at temperatures upwards of 250°C or under the influence of ultraviolet light, most of them involve bond homolysis, either in the backbone or in side groups attached to the backbone. For example, the first step in the thermal degradation of polystyrene is backbone homolysis:

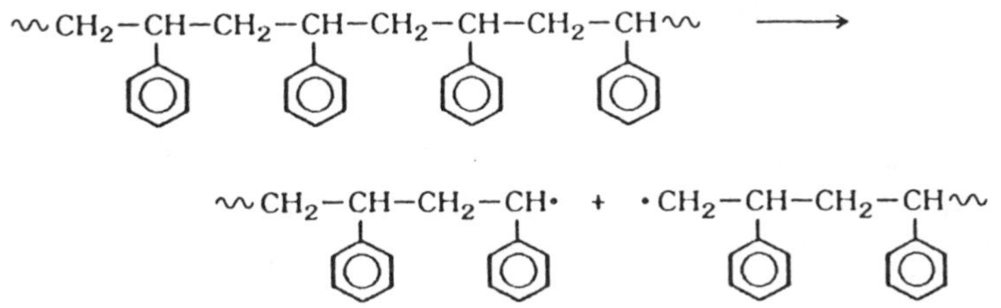

This is followed by various reactions of the terminal macroradicals so generated.

Homolysis in the side group occurs in the photolysis of poly(methyl acrylate) at 25°C, to give small amounts of volatile products; the macro-radicals so formed tend to form crosslinks:

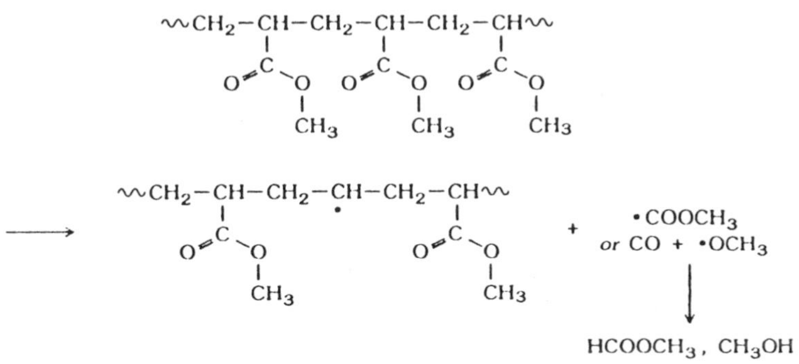

Some degradation reactions, however, are non-homolytic. There are various examples of concerted processes involving a six-membered ring transition state, as in the case of the photolysis of poly(ethylene terephthalate), where in addition to homolytic reactions, the following process, known as a Norrish type-II reaction, is thought to occur:

In other situations, notably hydrolytic and bio-degradation, ionic mechanisms are likely to be involved.

THERMAL DEGRADATION

Over the past forty years, the thermal degradation of a great variety of homopolymers and copolymers has been investigated. The topic has recently been reviewed by the author.[1] Some general patterns of behaviour have been established which can to some extent be related to the polymer structure. There are two main types of degradation process in the absence of air: when air is present, additional reactions may be involved.

Depolymerisation

This name can be applied to degradation processes in which the chain breaks at some point, leading to reactions in which the products all have essentially the same composition as the repeat structure, but may consist of relatively small molecules such as monomer, dimer and trimer, or chain fragments which are similar to the original polymer or copolymer structure, but of much shorter chain length. The simplest reaction of this type involves chain homolysis, followed by depropagation, as in the case of poly(methyl methacrylate):

monomer

In PMMA, the depropagation proceeds rapidly along the chain once the initial break has occurred, giving up to about 200 monomer units per initial scission. Monomer is the only thermal degradation product from this polymer, obtained in up to 100% yield.

Polystyrene also gives large amounts of monomer, by depropagation of radicals formed as shown previously. In this case, however, in addition to monomer, some dimer and trimer and a substantial amount of chain fragments are also produced. These result from transfer reactions of the radicals, involving abstraction of the reactive tertiary hydrogen atoms.

Polyethylene and polypropylene give only small amounts of monomer; transfer reactions produce a range of chain fragments covering a wide range of sizes.

The degradation of various heterochain polymers can be described broadly as depolymerisation. Polydimethylsiloxane breaks down simply to a mixture of cyclic siloxanes, but for most heterochain polymers depolymerisation to chain fragments is accompanied by side reactions of other types.

Depolymerisation processes, if not accompanied by other types of reaction, leave no ultimate solid residue of degradation.

Side Group Reactions

The fact that repeat units are held in proximity available for reaction between suitable functional groups has already been noted. Two types of reaction are commonly observed, side group scission or cyclisation involving adjacent repeat units. For example, PVC and poly(vinyl acetate) each lose a molecule of acid per original repeat unit:

$$\sim\sim CH_2-CH-CH_2-CH-CH_2-CH-CH_2-CH-CH_2-CH\sim\sim \longrightarrow$$
$$\underset{(OCOCH_3)}{\overset{|}{Cl}}\quad\underset{(OCOCH_3)}{\overset{|}{Cl}}\quad\underset{(OCOCH_3)}{\overset{|}{Cl}}\quad\underset{(OCOCH_3)}{\overset{|}{Cl}}\quad\underset{(OCOCH_3)}{\overset{|}{Cl}}$$

$$\sim\sim CH_2-CH-CH_2-CH-CH_2-CH-CH_2-CH-CH=CH\sim\sim$$
$$\underset{(OCOCH_3)}{\overset{|}{Cl}}\quad\underset{(OCOCH_3)}{\overset{|}{Cl}}\quad\underset{(OCOCH_3)}{\overset{|}{Cl}}\quad\underset{(OCOCH_3)}{\overset{|}{Cl}}\qquad + \quad HCl$$
$$(CH_3COOH)$$

Once an acid molecule has been lost, the double bond so formed destabilises the next repeat unit, so that the reaction tends to proceed along the chain, which results in the formation of a conjugated polyene:

$$\sim\sim CH=CH-CH=CH-CH=CH-CH=CH-CH=CH\sim\sim$$

This is responsible for the observed discoloration of each of these polymers. The product acid can catalyse the reaction. Furthermore, in the case of PVC, the initial loss of acid occurs at structural abnormalities. These features account for the unexpectedly low stability of PVC.

Polymers with COOH side groups dehydrate intramolecularly by inter-unit cyclisation. For example, in the case of poly(acrylic acid), adjacent pairs of units produce a cyclic anhydride structure:

$$\text{\textasciitilde CH}_2-\ \underset{\underset{O}{\overset{C}{\underset{|}{}}}{\underset{\text{OH}}{}}{CH}-CH_2-\underset{\underset{HO}{\overset{C}{\underset{|}{}}}{}}{CH}-CH_2-\underset{\underset{O}{\overset{C}{\underset{|}{}}}{\underset{OH}{}}}{CH}\text{\textasciitilde} \longrightarrow \text{\textasciitilde CH}_2-CH \overset{CH_2}{\underset{}{}} CH-CH_2-CH\text{\textasciitilde}$$

+ H₂O

There are many other examples of inter-unit cyclisations in carbon chain polymers, as for example in polyacrylonitrile, polybutadiene, polyisoprene, vinyl ketone polymers and various copolymers. Such reactions are not normally found, however, in heterochain polymers.

Three important features of side group reactions should be noted. When they can occur, they will often take place at lower temperatures than backbone scission and will compete with or prevent the latter. Since they often involve release of small molecule fragments, the products of degradation in such cases cannot be similar in composition to the repeat structure. Degradation may lead ultimately to a char residue.

The products of thermal degradation of some important carbon chain polymers are listed in Table 2.

Table 2. **Thermal Degradation Products from Some Addition Polymers, heated slowly up to 500°C.**

Polymer	Degradation Products
polyethylene, polypropylene	chain fragments of various sizes, including small amounts of volatile saturated and unsaturated hydrocarbons
polystyrene	about 50% yield of styrene monomer, plus some dimer and short chain fragments
natural rubber	isoprene, dipentene and short chain fragments including some cyclised structures
poly(vinyl chloride)	quantitative yield of hydrogen chloride; small amounts of benzene; coloured tars; carbonaceous residue
poly(vinyl acetate)	quantitative yield of acetic acid; small amounts of benzene; coloured tars; carbonaceous residue
poly(vinyl alcohol)	water; coloured tars; carbonaceous residue
poly(methyl methacrylate)	quantitative yield of monomer
poly(methyl acrylate)	some methanol and carbon dioxide; main products are short chain fragments.

An extreme, but important, case of thermal degradation is the situation in a fire. Strictly, polymers do not burn, because they cannot mix with air – they decompose to small molecules which with air may form a combustible mixture. The burning cycle for a polymer is illustrated in Fig. 5. To make a polymer safer from the point of view of flammability, it is necessary to break this cycle in one or more ways. This can be done at each of the points A, B and C, using fire retardant additives. The production of fuel may be impeded by a "condensed phase" effect in which the degradation reaction is modified (A), a flame quencher may be produced on heating (B) or a char or glassy layer may impede heat transfer (C).

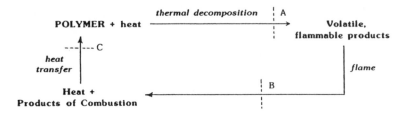

Fig. 5. Polymers in a Fire Situation: The Burning Cycle

PHOTODEGRADATION

The optical radiation spectrum (Fig. 6) extends from the far ultraviolet through the visible region into the near infrared. In terms of wavelength of radiation, the span is from 200 nm to about 1400 nm. The energy, E, associated with a photon of wavelength, λ, is related to Planck's constant, h, and the velocity of light, c, by the equation $E=hc/\lambda$, so that a corresponding scale of energies can be calculated. After conversion into energy per mole of photons, the values shown on Fig. 6 are obtained.

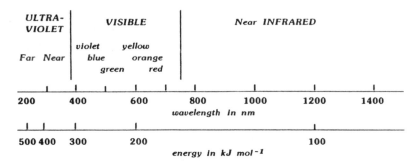

Fig. 6. Optical Radiation Spectrum

The bonds in macromolecules typically have energies between 300 and 500 kJ per mole. Visible and infrared radiation is of insufficient energy to break such bonds, but it can be seen that ultraviolet light with wavelength lower than about 400 nm is of suitable energy. Some of the sun's radiation, however, is absorbed in the atmosphere of the earth. The solar UV spectrum in summer at the earth's surface between 280 and 400 nm is shown in Fig. 7. and the notable feature is the cut-off at 290 nm (400 kJ per mole).

This means, in particular, that there is insufficient energy to break C—H or C—F bonds, but that suitable C—C, C—O and C—Cl bonds could be broken.

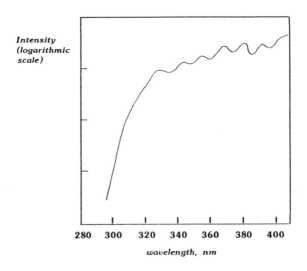

Fig. 7. **Ultraviolet Component of Summer Sunlight Reaching the Earth's Surface**

In order for UV light to cause photolysis of a polymer, it must first be absorbed. This requires a *chromophore*. This may be a group within the polymer structure, of which the most important examples are structures containing the carbonyl group. An alternative pathway in photodegradation, however, involves absorption of light by an additive and energy transfer to the polymer, with the formation of an excited polymer molecule which undergoes homolysis. The radicals so formed can then undergo various possible reactions. The basic processes in photolysis, excluding the effects of oxygen, may be summarised as follows:

1.   Absorption of the incident radiation

    a. by polymer
    b. by additive/ impurity } ⟶ excited molecule $P \longrightarrow P^*$
                                                        $S \longrightarrow S^*$

2.   Sensitisation (energy transfer)

$$S^* + P \longrightarrow S + P^*$$

3.   Homolysis of the excited polymer molecule (chain or side group scission)

$$P^* \longrightarrow R_a{}^{\bullet} + R_b{}^{\bullet}$$

where $R_a{}^{\bullet}$ and $R_b{}^{\bullet}$ are radicals, both or one of which is a macroradical.

4. *Macroradical reactions*

   a. disproportionation
   b. chain scission
   c. chain scission following intermolecular transfer
   d. depropagation to monomer
   e. crosslinking
   f. other processes.

Several polymers which do not contain a chromophore in the repeat structure are nevertheless found to be unstable in use in sunlight, unless protected by added photostabilisers. These polymers include PVC, polyethylene, polypropylene, natural rubber, polyamides and polypeptides. The explanation is the presence in the chains of some carbonyl groups formed as a result of oxidation.

Which of the several reaction possibilities under heading 4 occurs in the case of a particular polymer depends primarily on the glass transition temperature of the polymer and the chemical structure, in particular the presence or absence of tertiary hydrogen atoms.

Thermal degradation in most cases occurs when the polymer is in the form of a melt, so that any small molecule products of degradation can escape into the gas phase with relative ease. Such escape is possible, but less easy, when a polymer is subjected to photolysis above T_g. Below, T_g, however, depropagation to monomer, which is an important degradation reaction in some polymers at elevated temperatures, cannot occur following chain scission induced photolytically, because when the polymer is in the glassy state the propagation - depropagation equilibrium cannot be driven to the right by the removal of monomer:

$$R_n{}^\bullet \rightleftharpoons R_{n-1}^\bullet + M$$

Most polymers which photodegrade at normal temperatures will be below T_g, important exceptions being rubbers, polyethylene and polypropylene; these are polymers which in thermal degradation have been found to give little monomer as a product. In general, it is found that photodegradation of polymers at normal temperatures leads to only small amounts of any volatile products formed.

Among the reactions included under heading 4f are those leading to discoloration. Yellowing is a common effect of photodegradation and is due to the formation of conjugated structures, in some cases polyenes and in others double bonds conjugated to carbonyl or aromatic structures. In the case of PVC, its origin is similar to that in the case of thermal degradation although the reaction is limited by the fact that at normal temperatures the polymer is below its T_g.

In the case of carbon chain polymers, an important generalisation (to which there are a few notable exceptions) is that for polymers of repeat structure $\sim CH_2CXY\sim$, in which there are no tertiary hydrogen atoms, the observed predominant effect of photolysis by ultraviolet light is a fall in molecular weight due to chain scission, whereas those of repeat structure $\sim\!\!\sim CH_2CHX\sim\!\!\sim$ tend to crosslink. In each case it is assumed that a suitable chromophore is present. Tertiary hydrogen abstraction in poly(methyl acrylate), for example, leads to the formation of a backbone macroradical of the following structure:

$$\sim\!\!\sim CH_2-CH-CH_2-CH-CH_2-\overset{\displaystyle\cdot}{C}-CH_2-C\sim\!\!\sim$$
$$\underset{\displaystyle COOCH_3}{|}\quad\underset{\displaystyle COOCH_3}{|}\quad\underset{\displaystyle COOCH_3}{|}\quad\underset{\displaystyle COOCH_3}{|}$$

and two such radicals in different chains can form a crosslink.

These differences have a profound effect on the physical properties. Crosslinked polymers lose any flexibility and become insoluble; when chain scission occurs, physical properties deteriorate sharply and solubility is increased.

Some non-homolytic reactions can result in certain cases from absorption of ultraviolet radiation. One such reaction in poly(ethylene terephthalate) has already been noted.

ATMOSPHERIC DEGRADATION

Oxygen

The ground state of the oxygen molecule is a triplet state (denoted $^3O_2$) in which there are two unpaired electrons. Oxygen usually participates in degradation reactions of polymers as a free radical species. Although oxidative degradation can occur at normal temperatures and in the absence of UV light, the commonest effects result from the combined effect of oxidation and thermal or photodegradation. The oxidative chain mechanism is of prime importance, whenever there is a source of free radicals:

$$R\cdot + O_2 \longrightarrow RO_2\cdot$$
$$RO_2\cdot + RH \longrightarrow ROOH + R\cdot$$

In the second step, RH can be a suitable polymer molecule. Because of the chain nature of this reaction, even small concentrations of free radicals can result in significant amounts of oxidative degradation. The primary oxidation product, the hydroperoxide ROOH, is thermally and photolytically unstable. It decomposes to give two radicals, each of which can participate as $R\cdot$ in the chain process:

$$ROOH \longrightarrow RO\cdot + \cdot OH$$

It is a side reaction of the alkoxy radical which is of critical importance in relation to loss of physical properties. For example, in the case of polystyrene:

i.e. chain scission occurs, generating a macroradical which continues the chain process.

Polymers with tertiary hydrogen atoms on the backbone, or with methylene groups or methine groups activated by unsaturation, are particularly sensitive to oxidative degradation.

A second type of degradation due to oxygen arises when, in the presence of a suitable photosensitiser (S) capable of efficiently absorbing ultraviolet light, energy transfer occurs to produce an excited state of the oxygen molecule, singlet oxygen ($^1O_2$):

$$S \longrightarrow S^*$$
$$S^* + {}^3O_2 \longrightarrow S + {}^1O_2$$

For example, tyrosine and tryptophane amino acid residues present in the protein structure of wool can act as photosensitisers. The singlet oxygen formed then attacks the tryptophan structure to form a hydroperoxide. Singlet oxygen is unreactive towards saturated hydrocarbon chains, but will attack olefinic groups e.g. in rubber.

Nitrogen Dioxide, Sulphur Dioxide and Ozone

The oxides NO_2 and SO_2 are important constituents of atmospheric pollution. In moist conditions, the acids formed can accelerate hydrolytic attack on suitable polymers. These oxides can also add directly to olefinic double bonds to give products which undergo degradation. A further mode of action of NO_2 is as a photosensitiser for the formation of singlet oxygen, according to the mechanism shown previously.

Polyamides and polyurethanes are sensitive to attack by NO_2. Both chain scission and crosslinking result, although the former predominates. Initial attack is at the NH group:

$$-\overset{\overset{\text{O}}{\|}}{\text{C}}-\overset{\overset{\text{H}}{|}}{\text{N}}- \; + \; NO_2 \longrightarrow \; -\overset{\overset{\text{O}}{\|}}{\text{C}}-\overset{\overset{\bullet}{}}{\text{N}}- \; + \; HNO_2$$

$$\longrightarrow \text{ chain scission } \quad or \quad -\overset{\overset{\text{O}}{\|}}{\text{C}}-\overset{\overset{\text{NO}_2}{|}}{\text{N}}-$$

Although ozone is present in the atmosphere in low concentration, it is an effective pro-degradant for certain polymer structures, e.g. polyethylene, polystyrene, rubber and polyamides. The addition to double bonds in rubber to give the ozonide is followed by scission of the backbone:

$$\wedge\!\wedge CH_2-CH=CR-CH_2\wedge\!\wedge \longrightarrow \wedge\!\wedge CH_2-\overset{\overset{\displaystyle O^{\diagdown O\diagdown}O}{\diagdown \diagup}}{CH}-\overset{}{CR}-CH_2\wedge\!\wedge \longrightarrow \text{ scission}$$

Products are a shorter chain with an aldehydic end and either a polyperoxide or an iso-ozonide which can undergo further degradation reactions. The oxidised structures are effective chromophores for degradation by photo-oxidation.

With polyethylene, the reaction with ozone is much slower and is thought to proceed as follows:

$$\text{\tiny$\sim\!\!\sim$}CH_2-CH_2-CH_2\text{\tiny$\sim\!\!\sim$} \xrightarrow{O_3} \text{\tiny$\sim\!\!\sim$}CH_2-\overset{\overset{\displaystyle O-O\cdot}{|}}{CH}-CH_2\text{\tiny$\sim\!\!\sim$} + \cdot OH$$

$$\downarrow$$

$$\text{\tiny$\sim\!\!\sim$}CH_2-COOH + \cdot CH_2\text{\tiny$\sim\!\!\sim$} \quad or \quad \text{\tiny$\sim\!\!\sim$}CH_2-\overset{\overset{\displaystyle O}{\|}}{C}-CH_2\text{\tiny$\sim\!\!\sim$} + \cdot OH$$

$$or + polymer \longrightarrow \text{\tiny$\sim\!\!\sim$}CH_2-\overset{\overset{\displaystyle O-OH}{|}}{CH}-CH_2\text{\tiny$\sim\!\!\sim$} + \text{\tiny$\sim\!\!\sim$}CH_2-\overset{\displaystyle\cdot}{CH}-CH_2\text{\tiny$\sim\!\!\sim$}$$

Again, after initial attack by ozone, the oxidation chain can be established and chromophores for absorption of UV light are formed.

HYDROLYTIC AND BIODEGRADATION

Hydrolytic degradation is possible in synthetic polymers containing ester, amide, urethane and carbonate links and in natural polysaccharides and proteins. Apart from the case of polymers with ester and amide structures in the side groups rather than in the backbone, hydrolysis leads to a rapid loss of physical properties as a result of cleavage of the chains. Because of the hydrophobic character of most polymers, hyrolysis, even where feasible, may proceed slowly. Humid conditions and pH less than 7 will favour this type of degradation reaction.

Biodegradation of polysaccharides and proteins is induced by enzymes, which are complex proteins containing hydrophilic groups such as COOH, OH and NH_2. Enzymes are highly specific towards particular chemical structures.

Microorganisms may also be specific in their attack, but may be able to adapt to new substrates such as synthetic polymers. Polymer structures sensitive to attack by microorganisms are aliphatic polyesters, polyethers, polyurethanes and polyamides. Two commonly used types of polyurethanes are poly(ester urethanes) and poly(ether urethanes). The former are more susceptible to biodegradation than the latter.

The several types of microorganisms have different requirements for effective action. Fungi require oxygen and quite acid conditions, pH 4.5 - 5, and the optimum temperature is about 35°C. Actinomycetes and bacteria prefer less acid conditions, pH 5 - 7, and will operate over a wider temperature range, the optimum being about 60°C. Although actinomycetes require oxygen, bacteria can operate in aerobic or anaerobic conditions.

In addition to the functional group requirements noted above, it has been found that biodegradation is strongly influenced by chain length and branching. Short, linear chains are more susceptible. Polymers which are initially resistant to biodegradation may become susceptible after the chain size has been reduced by photo-oxidation.

Table 3. **Susceptibility of Unstabilised Polymers to Degradation**

POLYMER	Thermal Degradation	Photo-oxidation	Ozone	Hydrolysis	Bio-degradation
polyethylene	2	3	1	0	1
polypropylene	2	4	1	0	0
natural rubber	2	4	4	0	1
polystyrene	2	3	1	0	1
poly(vinyl chloride)	4	3	0	1	0
poly(vinyl acetate)	3	3	0	4	2
poly(vinyl alcohol)	4	1	0	0	3
poly(methyl acrylate)	2	3	0	1	1
poly(methyl methacrylate)	3	2	0	1	0
poly(ethylene terephthalate)	2	1	1	1	1
bisphenol A polycarbonate	1	3	2	1	0
polytetrafluoroethylene	0	0	0	0	0
polyamide (Nylon-6)	2	2	2	2	2
polyurethanes	3	3	1	1	2-4
polypeptides	2	2	2	2	4
alkyd resins	2	2	2	1	
epoxy resins	2	2	1	1	
cellulose	2	2	0	2	3

Key: Susceptibility to degradation *increasing* on the scale 0 to 4

In conclusion, the susceptibility of certain polymers to various types of degradation has been summarised in Table 3. Although this provides some useful comparisons, the data must be treated with caution because of the importance in degradation processes of sample history, in particular resulting in low concentrations of sensitive abnormal structures, and the uncertain effect of various additives or impurities.

Some useful reading sources are listed below.[1-5]

REFERENCES

1. I.C. McNeill, In *Comprehensive Polymer Science*, Vol. 6, ed. G. Eastmond, A. Ledwith, S. Russo, P. Sigwalt, Pergamon, London, 1989, Ch. 15.
2. N. Grassie and G. Scott, *Polymer Degradation and Stabilisation*, Cambridge, London, 1985.
3. W. Schnabel, *Polymer Degradation: Principles and Practical Applications*, Hanser, Vienna, 1981.
4. J.F. McKellar and N.S. Allen, *Photochemistry of Man-Made Polymers*, Elsevier Applied Science, London, 1979.
5. J.F. Rabek, *Mechanisms of Photophysical Processes and Photochemical Reactions in Polymers*, Wiley, New York, 1987.

Preservation of Natural Macromolecules

C. V. Horie

THE MANCHESTER MUSEUM, THE UNIVERSITY, MANCHESTER M13 9PL, UK

Introduction and background

Over the past three and a half thousand million years, a broad range of organisms have evolved. Their structures and processes are based on a common core of carbon chemistry which produces macromolecules of high complexity and specificity. The mechanisms by which these polymers are formed has been increasingly understood during recent decades[1], though these mechanisms cannot be duplicated by purely synthetic methods. Many of the molecules themselves have not been described. The only source of most natural macromolecules is the organism that creates them. If we wish to have a continued supply of these materials, it is necessary to ensure that the organisms reproduce themselves as a renewable resource. Nature conservation (not the subject of this paper) is concerned with the maintaining of viable populations of living organisms, ideally in the original environment which enabled the diverse species to develop. Increasingly, preservation of endangered species is being carried out in artificial protected areas such as zoos and botanical gardens. For various reasons, collections of live organisms are unsuitable for many purposes in the study of organisms, particularly taxonomy- the classification of organisms. The large collections in museums and comparable institutions were built up primarily to provide both the raw material and the research results for taxonomy, though the information gathered is used for a wide variety of other purposes.

What are the features or components in the preserved specimens that can be used to gain information or insights? How are these best preserved?

These questions had, and still have, a profound effect on the methods of researchers in natural history. In the distant past, from the ancient Greeks to around the sixteenth century AD, organisms were studied and their features recorded in words. Those descriptions which were written down and survived became the received knowledge of the age. This proved unsatisfactory. A description of an object is

Figure 1.   Head of the Dodo originally preserved at the Ashmolean.
Photograph kindly supplied by the Oxford University Museum (Zoological
Collections).

no substitute for the real thing. If a description is to be
reassessed or enlarged upon, the object itself must be
re-examined. For instance, records of mermaids have not been
substantiated by voucher specimens held in museums.   Thus
specimens must be preserved and kept for extended periods.
The growth of scientific curiosity in the Renaissance
developed alongside a wider fascination with the unusual.
Rich patrons gathered cabinets of curiosities while considered collections were prepared by scholars. More subtle
study of the natural world could be attempted only by
comparing similar specimens. Re-evaluation of animals and
plants by direct observation replaced the use of precedent.
During the early sixteenth century, an herbarium was
established in Padua by Luca Ghini, and a wider natural
history collection by Konrad von Gesner in Germany[2].
Through the seventeenth and eighteenth centuries, prolifer-
ation of collections[3] and classification of specimens
developed hand in hand, each collection largely the result
of the efforts of a single individual. Portions of some
collections survive. Much does not. The most notable
example is the dodo. A living bird was brought to England
in 1638, probably then being preserved as a stuffed specimen
in the Ashmolean Museum. After a hundred years of deterioration, the mount disintegrated, only a few pieces being
rescued, Figure 1.

Acquisition of specimens for taxonomic purposes accelerated
during the nineteenth century, Figure 2. Many collections
were deposited in museums pleased to take part in the then

fashionable and exciting science. Unfortunately this enthusiasm failed in many cases to ensure adequate care. A report on collections in the UK[4] exposed the large number of collections that either had been destroyed as a result of poor curation over the past hundred years, or were actively deterio- rating. Many collec- tions of importance are just unrecognised. The old specimens retain their importance. Each bio- logical species name is defined by the name

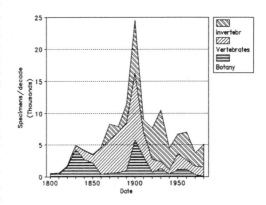

Figure 2. Number of plant and animal specimens collected per decade, from a partial survey of five UK museums.

given to a single specimen, a *type specimen*. These and closely associated specimens are in considerable demand when identifying or naming similar organisms. Many of the type specimens are the oldest ones of that species to be pre- served.

Methods by which specimens are preserved change as the requirements of the research change. Until the last two decades, research on museum specimens has largely been carried out by studying morphology, the shape and inter- relationship of physical form. Work started on the large or easily accessible specimens like mammals and flowering plants and progressed to microscopic animals, plants etc. These were preserved as dead specimens, whole or in part. Some primitive organisms such as fungae, algae and bacteria cannot be distinguished when killed and preserved. For this reason, their taxonomic collections are maintained as living cultures.

Up to the late seventeenth century, long term preservation was possible only of dried specimens. For instance, animals frequently had much of the soft tissue removed. The introduction of ethanol solutions, "spirits of wine", rum etc., as a fixative and preservative in the 1660s, enabled soft tissue to be kept in a state closely approaching that of the original[3]. The technique was rapidly adopted by naturalists and anatomists and its use continues to this day. Other preservatives and many modifications, some highly idiosyncratic, to the ethanol preservative were tried in succeeding centuries. The next introduction of continu- ing importance was that of methanal (formaldehyde) solutions by Blum in the 1890s, as a fixative for animal tissue[5]. Freeze-drying[6] has enabled the preservation of tissue in a dry state that must otherwise have been discarded. Increas- ingly, specimens and samples are being held in the frozen state[7].

Each of these methods, drying (with evisceration), dewatering with ethanol, fixing with methanal, freeze-drying and freezing causes considerable changes in the specimen. However, researchers have learned to discount the physical changes caused by preparation and preservation techniques. Even so, ambiguities have arisen when colour changes are caused by deliberate or accidental chemical treatments[8,9].

Increasingly, specimens are being examined at a scale where the molecular architecture becomes important[10]. For instance the phylogeny of birds was examined by analysing the pigment types and components in feathers[11] taken in part from museum specimens, this being the obvious and convenient way to assemble a suitable range of bird species. However, it is well established for (mammal) hair that the keratin changes over time[12], both *in vivo* and *in vitro*. The nature and rate of change in the molecular components need to be understood if comparisons between preserved specimens of differing histories are to be related to their living states. Chemical methods of examining specimens will become increasingly important[13]. However, morphological methods will probably remain dominant for taxonomic studies, partly because these methods are cheaper and less technically complicated, but mostly because specimens must be compared with *types* which have been described in this way. Methods of reliably preserving both the shape and chemical components must be established.

Initial stages, and enzymes

Living organisms are composed of a number of diverse components[14], Figure 3. The largest component of most organisms is water, though the proportion in different components varies from organ to organ. In living organisms, components are elaborated, laid down and degraded in water. When water is excluded from a portion of an organism, that part no longer takes part in the living process and is effectively dead. Examples are the lignification of cellulose which creates a hydrophobic matrix and thus

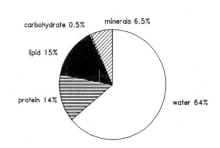

Mammal body

carbohydrate 0.5%   minerals 6.5%

lipid 15%

protein 14%   water 64%

Figure 3. Composition of the typical human body, data from ref. 14

protective encapsulation for the cellulose, and the filling by keratin of epidermal cells which die and act as sacrificial protective layers for the underlying living tissue. Living process make use of enzymes which are highly specific catalysts. A suite of cooperating enzymes carry out the chemical reactions that together are called life. For instance, some enzymes make collagen in skin, others break it down when the skin must be repaired and renewed. The control of the relative activity of the competing activities requires constant monitoring and control. This takes energy.

When the organism or cell dies, the enzymes continue to operate. In general, constructive reactions require an input of energy which will eventually be exhausted. However, this can take some time. A flower in a vase can bloom for days using stored energy.

Muscle maintains its relaxed configuration by ATP (adenosine triphosphate), whose concentration is kept at high levels by oxygen supplied through the blood[1]. When blood circulation fails, ATP is hydrolysed by enzymes. The muscle goes rigid, and contracts, producing *rigor mortis* after a few hours causing distortion. Subsequently, the muscle relaxes when proteases (enzymes acting on proteins) degrade the proteins holding the muscle fibres in place. This causes tendering of meat. Similar degradative reactions occur in skin. If left too long after removal from the carcass, endogenous enzymes can cause the epidermis to slip and hair to fall out[15]. These enzymes are active in more or less physiological conditions.

Some enzymes continue active even in nominally "dry" conditions. For instance, ATPase the enzyme that breaks down ATP has been studied[16]. Freeze-dried meat was stored at various relative humidities. Between 25 and 40%RH, the ATPase became active and continued hydrolysing ATP, even in these dry conditions. When meat stored at 25%RH for 3 months and showing little loss of ATP was wetted, the inactive ATPase became active though at a lower level than when fresh. Other enzymes acting on different substrates show similar behaviour; e.g. α-amylase which hydrolyses starch at 53%RH and a pork extract enzyme which degrades glycogen below 11%RH[17]. Some lipases, fatty acid releasing enzymes, have been shown to react with lipids at very low humidity levels, ca 3%RH. These lipase reactions occur in animal but especially plant material where the cell structure has been disrupted. The living cell contains these enzymes in separate compartments from its lipid substrate. In general, high relative humidity levels, ca.50%RH, are needed to enable the enzymes to act on water sensitive substrates. The limiting factor is the need to mobilise the molecules sufficiently.

Over time, enzymes gradually denature and lose their activity. However, changes must be expected for some years. The aim of many preservative treatments is to denature the degradative enzymes, though this of course causes the loss of chemical information in both these and other sensitive enzymes. Unfortunately, freezing a specimen does not necessarily stop enzymatic reactions. Indeed some are speeded up because the enzyme and its substrate are concentrated in the liquid solution remaining when much of the water is removed as ice crystals. Enzymatic action can continue at temperatures as low as -23°C, though activity is gradually lost[18].

Non-Enzymatic browning

Another, possibly short-term, cause of change is due to reactions of reducing sugars with themselves and proteins,

$$
\begin{array}{lll}
\text{HC}=\text{O} + \text{H}_2\text{N--R} & \text{HC--NH--R} & \text{H}_2\text{C--NH--R} \\
| & \| & | \\
\text{HC--OH} & \text{C--OH} & \text{C}=\text{O} \\
| & | & | \\
\text{HO--CH} & \text{HO--CH} & \text{HO--CH} \\
| & | & |
\end{array}
$$

1−amino−1−deoxy−2−ketose

Amadori compound

the Maillard reaction[19], The aldose group on the sugar molecule reacts with a free amine group to form a cross-link. These reactions can and do occur in the absence of oxygen. The products are highly reactive and progressively isomerise and condense. Volatile products produce off-odours, perhaps the source of museum mustiness. Non-volatile products are coloured, usually brown but sometimes brighter, and can contain free radicals.

One of the major amine groups involved in the reaction is the ω-amino group of lysine amino acid residues in proteins, which makes up about 6% of leaf protein and 10% of animal protein. Protein is cross-linked by reaction with the sugars[20], so making it unavailable for study and reducing enzymatic activity. This will also cause changes to physical properties and colour.

$$
\begin{array}{c}
\text{NH}_2 \\
| \\
\text{CH}_2 \\
| \\
\text{CH}_2 \\
| \\
\text{CH}_2 \\
| \\
\text{CH}_2 \\
| \\
\text{---NH---CH---CO---}
\end{array}
$$

Lysine

The rate and activation energy of the reaction is highly dependent on water activity[21], with a maximum rate at around 50-70%RH, Figure 4. At low water activities the activation energy, normally around 20-25kcal/mole, rises to around 40kcal/mole. The reaction depends on the presence and mobility of the components. The initial reaction between the available sugar and amino groups probably goes to completion within a few months[22], though further reactions will continue for some time. Similar reactions occur between sugars and DNA causing insolubility, with faster reaction occurring with denatured DNA[23].

Water relationships

Water is held by organic polymers in a hierarchy of strengths. The most closely bound water is that which forms part of the structure of the polymer. In the case of collagen, this is about two molecules per tripeptide unit forming part of the triple helix, i.e. about 0.07g water/g dry collagen, at an equilibrium humidity around 10%RH[24].

Between about 0.07 to
0.25g water/g collagen (up
to ca. 70%RH), water is
bound between the collagen
polymer structure, pushing
the molecules apart and
reducing its rigidity by
plasticisation. This
water is mobile enough to
take part in chemical
reactions, though it is
not freezable and is thus
not present as liquid
water. Removal of this
water by desiccation
forces the molecules close
together, and is likely to
give rise to irreversible
reactions. Above this
level, the water molecules
accumulate as liquid
water.

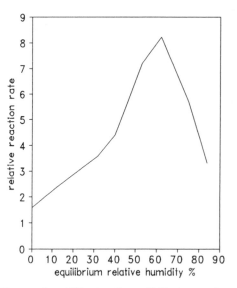

Figure 4. Effect of equilibrium relative humidity on browning in freeze-
dried pork. Data taken from ref 21.

This three stage process
of water absorption is
typical of organic
materials with both highly organised and amorphous components.

Studies in various fields have shown that the storage of
organic material alters their response to relative humidity
changes. Dried materials exhibit a hysteresis during a
cycle of sorption/desorption of water. This reflects the
relative accessibility of hydrophilic groups within the
structure. On ageing, polysaccharides show relatively
little change in the course of the hysteresis loop, reflect-
ing a fairly stable set of hydrophilic groups in the
structure. However, highly protein materials demonstrate an
overall decrease in the ability to sorb water, Figure 5,
combined with a progressive increase of the area of the
hysteresis loop[25]. It seems likely that groups which take
part in hydrogen bonding gradually align themselves together
so reducing their interaction with water. The polymer will
thus have reduced hydrophilicity. The change in response
to water vapour of protein is seen both in cooked meat[26]
and in skin[27] after storage. Collagen as initially laid
down is lightly cross-linked, Figure 9. Over decades of
ageing in skin, the molecules are further cross-linked by an
oxidative reaction both *in vivo* and *in vitro* with mammal
collagen[28]. Though the reaction does not occur in living
fish whose collagen is turned over regularly[29], the crosslinking reaction might take place in preserved fish. This
cross-linking causes the collagen to be stiffer and less
responsive to changes in relative humidity.

Skin is a composite of collagen and polysaccharide chains.
Over time, preserved skin will change its behaviour in

response to water content changes. The polysaccharide behaviour would remain roughly the same while the protein component behaviour would become more extreme to fluctuations. Therefore with ageing, proteinaceous material should receive more careful relative humidity control. At low water contents, the polysaccharides will be stiff because of multiple inter- and intra-molecular hydrogen bonds. Addition of water reduces the danger of stress concentrations between protein and polysaccharide components. The stress generated between adjacent materials can be on a macro scale, e.g. a wooden framework for a mounted skin, or on a micro scale, as between adjacent molecules. The former leads to gross distortions, the latter will lead to fragility.

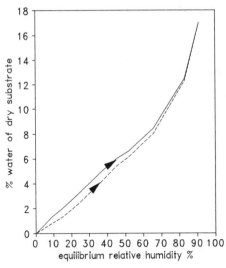

Figure 5. Change in absorption by high protein freeze-dried meat. Extrapolation from results given in ref. 25. _____ indicates the absorption in newly prepared material. _ _ _ _ indicates the behaviour after storage for 6 months at 100°F.

Lipid oxidation

	fatty acid	R
	palmitic (C_{16})	$CH_3(CH_2)_{14}-$
	oleic (C_{18})	$CH_3(CH_2)_6CH_2 \overset{CH=CH}{\diagup} CH_2(CH_2)_6-$
	linoleic (C_{18})	$CH_3(CH_2)_3CH_2 \overset{CH=CH}{\diagup} CH_2 \overset{CH=CH}{\diagup} (CH_2)_7-$
	arachidonic (C_{20})	$CH_3(CH_2)_3 \left[CH_2 \overset{CH=CH}{\diagup} \right]_4 (CH_2)_3-$

Triacylglycerol

Lipids are present in various forms in organisms[30]. The most obvious are fats and oils, composed of triacylglycerols where three fatty acids are combined with a glycerol molecule. The important distinction here is the number of double bonds in each of the constituent fatty acids. Lipids containing predominantly saturated fatty acids will be solid at room temperature, e.g. mammal fats, while unsaturated fatty acids create liquids or semi-solids, e.g. vegetable oils. Animals with low body temperatures such as fish contain a higher proportion of unsaturated acids. Less important on a weight basis but crucial to the structure of cell walls are the phosphoglycerides, a family of esters of glycerol with fatty acids and phosphoric acid.

The fatty acid is more available for reaction than when combined in the triacylglycerol. Fatty acids are released by enzymes, lipases. Both free and combined, the fatty acids oxidise readily in air. Oxidation is most vigorous for highly unsaturated molecules. The oxidation reaction is fast where there is free access to oxygen. When this is restricted, for instance in freeze-dried systems and in naturally dried oil storage cells, the lipids are encapsulated in membranes of low oxygen permeability. Water frequently acts as a plasticiser for these membranes, allowing oxygen to penetrate and the oxidation process to initiate[31].

The oxidation of lipids occurs both enzymatically and by simple autoxidation. Enzymatic oxidation and production of peroxides occurs in frozen plant material[18], resulting in the production of unsaturated aldehydes, and other changes at temperatures as low as −23°C in nitrogen. These reactions can be prevented by heating the material in order to denature the enzyme. Autoxidation of lipids[32] occurs through a well established peroxidation route which results in the production of free radicals, cross-linking of the lipids and production

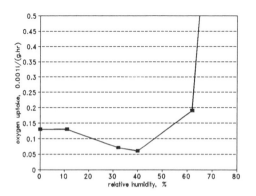

Figure 6.   Rate of oxygen uptake of potato crisps as affected by the ambient relative humidity. Data taken from ref. 34.

of lower molecular weight oily and volatile products. These reactions have low activation energies, below 20kcal/mole[33]. Lipid autoxidation reactions are also sensitive to water activity in the material[34], Figure 6. Storage in low humidity conditions will encourage this oxidation route.

Autoxidation is inhibited by anti-oxidants. Because of their importance, the cell is provided with an array of chemicals, such as vitamin E (retinol) and bilirubin, which provide protection at least in the short term. Some of the intermediaries in the Maillard reaction also have considerable anti-oxidant effect[35]. These two sets of anti-oxidants will be gradually consumed in dry preserved specimens as the lipids oxidise, contributing to the induction phase of autoxidation. Many zoological specimens appear stable for some years before sudden deterioration appears, in the form of "fat burn" of skins or greasy bones. Lipid oxidation is slowed by reducing the oxygen concentration of the ambient air below 10%[34].

The oxidation reaction has two effects. First, part of the lipid component is made more mobile and will migrate through the specimen to cause staining or worse. Second, the production of free radicals promote changes in the proteins of the specimen. In general, the globular proteins such as enzymes and albumins are cross-linked[36] while collagen is degraded[37]. Little significant incorporation of the lipid components in the protein occurs as the process proceeds by radical transfer. In skin, the reaction is apparent as loss of strength, denaturation and collapse.

What are the "right" conditions for storage of dry preserved specimens?

There are a number of potential reaction processes which affect the survival of the polymers present in the specimen. Their rates are in part determined by the external conditions of temperature, humidity and oxygen concentration. For dry specimens, loosely defined as those having no free water present, lowering the temperature might appear to reduce all rates of reaction. However for a complex system, it has been pointed out that reducing the rate of an antagonistic reaction (e.g. Maillard reaction) may allow another (e.g. lipid oxidation) to proceed faster[33]. Reducing the temperature of storage changes the moisture content of the specimen at equilibrium with a given relative humidity of the air.

On first examination, the effects of humidity on damaging reactions in storage lead to contradictory conclusions. No humidity range appears to provide overall reduced chemical deterioration. The relative importance of these reactions will change over time. The initial Maillard reactions go to completion within a fairly short time. Enzymes are sensitive to damage and so will be steadily denatured, by oxidation, the Maillard and lipid reactions. However, lipid oxidation has the potential to continue over a much longer term. Each molecule can react more than once. The oxidation initiated infects other components. If it is assumed that lipid oxidation is the most significant degradative process over the long term, humidity for storage should be around 40%RH.

Observations from long term ageing

The following observations are drawn primarily from the changes in protein materials[38]. Hair and the epidermal layer of skin are composed largely of keratin. This protein is characterised by large amounts of sulphur containing amino acid, cysteine, which creates S-S cross-links. With weathering, hair keratin evolves volatile sulphides[39] as a result of both oxidation and hydrolysis. The physical structure of the hair breaks down both on the intra- and inter-cellular level[40].

The reactions and results are different in low levels of water. When exposed to UV and oxygen, the pattern failure is different - the fracture surface shows no internal structure[41], Figure 7.   This implies that chemical reactions have so degraded the keratins that they have reduced in strength to that of the weakest component, or that cross-linking has occurred between the previously chemically and physically distinct structures.   A combination of both routes is likely.   The degradation mechanism affects the structural components and presumably also affects non-structural molecules that may be of interest in biochemical work.

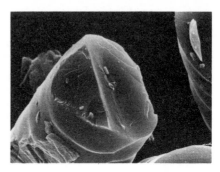

Figure 7.   Fracture surface of a hair exposed to UV and oxygen. Photograph by courtesy of Dr.I.L. Weatherall.

Collagen is the major component of skin and connective tissue in vertebrates, Figure 9[42,43,44].   The cross-linking which occurs soon after laying down are gradually supplemented by a slow oxidative cross-linking.   After a few decades, this process is complete *in vivo* and probably *in vitro*.   Collagen also degrades over time.   The first part to be lost is the non-helical ends of the molecules, so reducing the cross-linking and the stability of the structure.   The helical parts of the molecules are then attacked, causing further loss of stability, Figure 8.   The loss of stability can be determined by the reduction in the shrinkage temperature, T_s, when the collagen molecules unwind into the three monomers to become gelatine.   Hydrolysis of collagen occurs significantly if the ambient humidity is above 40%RH[45] while other reactions, presumably oxidative, predominate below 40%RH and in acid conditions. When T_s drops to near ambient temperature, the collagen denatures spontaneously.

Figure 8.   Degraded and partially collapsed collagen fibril from 2000 year old parchment.   Photograph courtesy of Dr.R.Reed.

For instance, it has been reported that the skin of Ginger, a 5000 year old Egyptian mummy, had shrunk badly and is soluble in water[46].

Both keratin and collagen are major components of vertebrate tissue.   However the significance of minor components has

D. Collagen Fibril

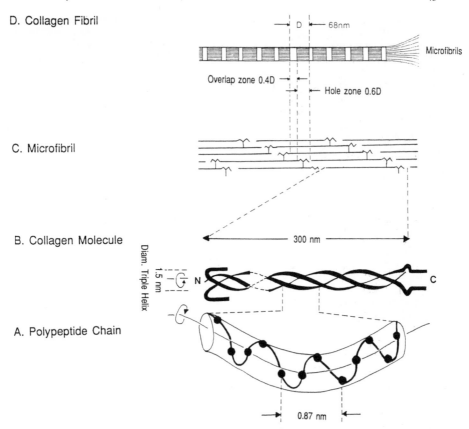

Figure 9. Lower orders of assembly of collagen Type I, from collagen monomer to fibril. The diagram has been redrawn from various sources.

been increasingly recognised. Of particular importance is DNA, deoxyribonucleic acid, the source of genetic code for organisms, present as large molecules in a cell. It is only in the past decade that the potential of museum specimens for preserving genetic information in this form has been realised. The DNA can provide valuable information about the genetic phylogenetic relationship of extinct organisms. Potentially, genetic characteristics may be transferable from preserved specimens into living organisms. A molecule of human DNA is very large, with millions of deoxyribonuc-liotide units making a molecule ca. 1m long. Even insect DNA is ca. 2mm long. It is extremely fragile and can rarely be extracted even from fresh tissues in other than fragmen-tary condition. DNA has been extracted from surprising specimens, for instance fossil Miocene magnolia[47], mummies[48] and extinct animals such as the thylacine[49]. The molecules appear to survive only as short lengths, a few hundreds of units long. Fortunately, multiple copies of these can be copied and multiplied by the polymerase chain reaction. This generates sufficient material to be sequenced and thus compared with other DNA. The process works if the surviving DNA chain has not been cross-linked or modified. Various techniques of museum preparation can make the DNA unavailable. Both arsenic and formaldehyde cross-link DNA while other tanning agents modify the chain, though DNA has been extracted from formaldehyde treated specimens, presumably because the membranes of the cell and nucleus have restricted access by formaldehyde to the DNA.

Conclusion

The aims for the preservation of museum specimens is changing as the questions being asked of them are informed by increasing knowledge and improved analytical techniques. Museums are not only custodians of the shape but also the material evidence, to the lowest level of structure. The processes of conservation must change to reflect this.

"An object is probably a write off .. when it ceases to have recognisable form." [50]

Even the goo which is all that remains of spirit specimen at the bottom of a jar may still contain irreplaceable information.

Acknowledgements

I am grateful for useful comments from C.W.Pettitt and to Dr.I.L. Weatherall for the use of SEM micrographs.

References

1.   L.Stryer, "Biochemistry" W.H.Freeman (1988)

2.   P.J.P.Whitehead, Museums in the history of zoology, Museums J. 1970 70 50

3.   F.J.Cole, "A History of Comparative Anatomy" Macmillan 1944

4.   B.Williams, "Biological Collections UK" Museums Association London, 1987

5.   H.F.Steedman, General and applied data on formaldehyde fixation and preservation of zooplankton, in "Zooplankton Fixation and Preservation", H.F.Steedman ed. Unesco 1976

6.   R.O.Hower, "Freeze-Drying Biological Specimens: A Laboratory Manual", Smithsonian Institution Press, 1979

7.   H.C.Dessauer & M.S.Hafner, "Collections of Frozen Tissues", Association of Systematics Collections, 1984

8.   E.R.Hall, Deleterious effects of preservatives on study specimens of mammals J.Mammalogy 1937, 18 359

9.   C.H.Fry, The affect of alcohol immersion on the plumage colours of bee-eaters, Bull.Brit. Ornith. Club, 1985, 105 78

10.   G.F.Barrowclough, Museum collections and molecular systematics, "Museum Collections: their roles and future in biological research", in E.H.Miller ed. British Columbia Provincial Museum, 1985, 43

11.   A.H.Brush & N.K.Johnson, Evolution of color differences between Nashville and Virginia warblers, Condor 1976, 78 412

12.   J.A.Maclaren & B.Milligan, "Wool Science, The Chemical Reactivity of the Wool Fibre", Science Press, 1981

13.   "Systematic Biology Research (written evidence received up to 21st May 1991)", Select Committee on Science and Technology (Sub-committee II) House of Lords paper 41, 1991

14.   E.S.West *et al* "Textbook of Biochemistry", Macmillan 1966

15.   B.M.Haines, Skin before Tannage - Procter's view and now, J.Soc. Leath.Technol.Chem. 1984 68 57

16.   K.Potthast, R.Hamm, & L.Acker, Enzymatic reactions in low moisture foods, in "Water Relations of Foods", R.B.Duckworth ed. Academic Press 1975, 365

17.   J.E.McKay, Behaviour of enzymes in systems of low water content, "Water Activity and Food Quality" T.M. Hardman ed. Elsevier 1989, 169

18.   W.D.Powrie, Chemical effects during storage of frozen foods J.Chem. Educ. 1984, 61 340

19.   F.Ledl, Chemical pathways of the Maillard reaction, in "The Maillard Reaction in Food Processing, Human Nutrition and Physiology" P.A.Finot, H.U.Aeschbacher, R.F.Hurrell & R.Liandon eds. Birkhäuser Verlag 1990, 19

20.   Y.Kato, T.Matsuda, N.Kato, & R.Nakamura, Maillard reaction in sugar protein systems, "The Maillard Reaction in Food Processing, Human Nutrition and Physiology", P.A.Finot, H.U.Aeschbacher, R.F.Hurrell & R.Liandon eds. Birkhäuser Verlag 1990, 97

21.   M.Karel, Chemical effects in food stored at room temperature, J.Chem.Educ. 1984, 61, 335

22.   K.Eichner, R.Laible, & W.Wolf, The influence of water content and temperature on the formation of Maillard reaction intermediates during drying of plant products, in "Properties of Water in Foods", D.Simatos & J.L.Multon eds. Martinus Nijhoff (1985) 191

23.   A.J.Lee & A.Cerami, Nonenzymatic reaction of reducing sugars with DNA, in "The Maillard Reaction in Food Processing, Human Nutrition and Physiology", P.A.Finot, H.U.Aeschbacher, R.F.Hurrell & R.Liandon eds. Birkhäuser Verlag 1990, 415

24.   S.Nomura, A.Hiltner, J.B.Lando, & E.Baer, Interaction of water with native collagen Biopolymers 1977, 16 231

25.   M.Wolf, J.E.Walker, & J.G.Kapsalis, Water vapour sorption hysteresis in dehydrated food, J.Agr.Food Chem. 1972, 30, 1073

26.   J.Strasser, Detection of quality changes in freeze-dried beef by measurement of the sorption isobar hysteresis, J.Food Sci. 1969, 34 18

27.   L.P.Witnauer, & W.Palm, Influence of cycling conditioning on the hydrothermal stability of leather, J.Amer.Leath.Chem.Assoc. 1968, 63 333

28.   B.J.Rigby, Aging patterns in collagen *in vivo* and *in vitro*, J.Soc.Cosmet.Chem., 1983, 34 439

29.   R.M.Love, "The Chemical Biology of Fishes" vol.2, Academic Press 1980

30.   M.I.Gurr, & A.T.James, "Lipid Biochemistry - An Introduction", Chapman & Hall 1980

31.   M.Karel, Effects of water activity and water content on mobility of food components, and their effects on phase transitions in food systems, in "Properties of Water in Foods", D.Simatos, & J.L. Multon, eds. Martinus Nijhoff 1985 153

32.   H.W.-S.Chan ed. "Autoxidation of Unsaturated Lipids", Academic Press 1987

33. M.Karel, Chemical effects in food stored at room temperature, J.Chem.Educ. 1984, 61 335

34. D.G.Quast & M.Karel, Effects of environmental factors on the oxidation of potato chips, J.Food Sci. 1972, 37 584

35. K.Eichner, Antioxidant effect of Maillard reaction intermediates, in "Antioxidation in Food and Biological Systems" M.G.Simic & M.Karel eds. Plenum Press, 1980, 367

36. J.Funes & M.Karel, Free radical polymerisation and lipid binding of lysozyme reacted with peroxidizing linoleic acid, Lipids 1981, 16 347

37. M.Karel & S.Yong, Autoxidation-induced reactions in foods, in "Water Activity: influences on food quality" L.B.Rockland, & G.F.Stewart eds. Academic Press (1981) 511-529

38. C.V.Horie, Deterioration of skin in museum collections, Polymer Degradation and Stability 1990, 29 109

39. J.H.Dusenbury, The molecular structure and chemical properties of wool, in "Wool handbook", vol.1, W.von Bergen ed. Interscience, New York, 3rd edn 1963, 211

40. W.Montagna & P.F.Parakkal, "The Structure and Function of Skin", Academic Press, New York, 3rd edn. 1974

41. I.L.Weatherall, The tendering of wool by light, in Proc.5th Int.Wool Text.Res.Conf., Aachen 1975 II, 1976, 580

42. S.Sakomoto, Bone, in "Collagen in Health and Disease", J.B.Weiss, & M.I.V.Jayson eds. Churchill Livingstone, Edinburgh 1982 362

43. J.A.Chapman, & D.J.S.Hulmes, Electron microscopy of the collagen fibril, in "Ultrastructure of the Connective Tissue Matrix", A.Ruggeri & P.M.Motta eds. Martinus Nijhoff, Boston 1984, 1

44. N.D.Light & A.J.Bailey, Covalent cross-links in collagen: Characterization and relationships to connective tissue disorders, in "Fibrous Proteins: Scientific, Industrial and Medical Aspects", vol.1, D.A.D.Parry & L.K.Creamer eds., Academic Press 1979, 151

45. J.A.Bowes & A.S.Raistrick, The action of heat and moisture on leather Part VI - Degradation of the collagen, J.Amer.Leath. Chem.Assoc., 1967, 62 240

46. C.Johnson & B.Wills, The conservation of two pre-dynastic Egyptian bodies, in "Conservation of Ancient Egyptian Materials", S.C. Watkins & C.E.Brown eds., United Kingdom Institute for Conservation, 1988, 79

47. E.M.Golenberg et al, Chloroplast DNA sequence for a Miocene Magnolia species, Nature 1990 344 656

48. S.Pääbo, DNA is preserved in ancient Egyptian mummies, in "Science in Egyptology", A.R.David ed., Manchester University Press, 1986, 383

49. R.H.Thomas, W.Schaffner, A.C.Wilson & S.Pääbo, DNA phylogeny of the extinct marsupial wolf, Nature 1989, 340 465

50. S.Bradley, Measuring deterioration: a finite life for objects?, in "Managing Conservation", Keene,S. ed., United Kingdom Institute for Conservation, 1990, 24

Oriental Lacquer: A Natural Polymer

H. F. Jaeschke

R & H JAESCHKE (CONSERVATORS), 3 PARK GARDENS, LYNTON, DEVON EX35 6DF, UK

1 INTRODUCTION

Oriental lacquer is a natural polymer that has been valued for millenia for its great durability and beauty. Excavations in Japan[1,2] and China[3,4] have revealed objects of wood and basketry with a red or black coating of oriental lacquer or a similar resin from prehistoric sites, possibly as early as 4,000 BC.

The lacquer is formed from a tree sap, obtained most commonly in China and Japan from Rhus verniciflua Stokes[5], though other regions have developed the use of similar substances from R. succedanea, R. ambigua, Melanorrhoea usitata, M. laccifera and Semecarpus vernicifera, among others. Variations in the properties of the lacquer formed from the sap are caused not only by difference in species, but also by difference in the age of the tree, the region in which it grows, the part of the tree sampled and the time of year.

This range in properties has been explored by craftsmen and different recipes and uses for various saps have become established. Some produce a higher gloss, some a greater clarity. Others are favoured for their hue, their durability or their speed of setting.

Further variations are obtained by mixing the treated sap with other constituents, such as ligroin, camphor or perilla oil to dilute it or to yield a higher gloss. The unpolymerised lacquer is extremely reactive with many pigments, limiting the range that can be used mainly to oxides and sulphides, but from the earliest times it has been coloured black by reaction with iron compounds (commonly iron acetate), or pigmented black with carbon or red with iron oxide or cinnabar. Lacquer is frequently mixed with powdered clay or starch paste to form a foundation layer, and the upper layers may be decorated with metal powders and flakes, sometimes mixed with charcoal.

A lacquer coating thick enough to be carved can be built up by applying many layers of the same or different coloured lacquer. Figure 1 shows a small wine cup formed from a silvered metal bowl to which a thick foundation layer of clay and treated sap was applied. This was covered with alternate thin coats of red and black lacquer which were then carved in imitation of a Chinese design. Figure 2 shows a thin-section of this work, revealing the foundation layer and above it dark bands of red lacquer with grains of cinnabar (mercury sulphide), interspersed with lighter transparent bands of black lacquer, coloured by reaction with iron salts, probably using the traditional method of treating iron filings with vinegar or rice wine and mixing the product into the treated sap. The figure also demonstrates the great limitations which black and white printing impose on light microscopy studies of lacquer.

Figure 1 Metal wine cup with carved black and red lacquer

Figure 2 Thin-section of black and red lacquer.

Throughout the centuries craftsmen have produced a vast range of lacquered objects, from minute decorative fittings to entire buildings, varying the techniques and materials involved to suit the properties of the lacquer used, the aesthetics of their culture and the nature of the finished article. Lacquer has been applied to wood and other organics, including bamboo, hide and paper, to bone and shell, to metals and even to ceramic. It bonds well to most substrates, producing a hard lustrous finish that protects the object from moisture, variations in temperature, many corrosive agents and even some physical damage. It is strong enough when set to be used with paper or clay powder to form "dry lacquer", three-dimensional objects with no core or substrate. Increasingly the actions of the craftmen in choosing and treating the sap are being revealed as skillful manipulations of a complex polymerisation process which is only gradually beginning to be understood.

2 PREPARATION AND USE OF THE RAW MATERIAL

The sap is a watery grey or milky liquid when first collected from cuts in the tree bark and may froth with the action of laccase, one of the enzymes present. An external aqueous phase is present, which separates out,

leaving the water in oil emulsion. After filtering to
remove larger particles and impurities, the sap is
stirred in an open pan, sometimes with gentle heat until
the correct turbidity is reached. The treated sap is then
stored, traditionally in small wooden tubs covered with
paper. Periodically these are opened and any further
froth removed. The treated sap may be used as soon as it
is prepared, but it is common to allow it to remain in
the tubs to season, often for a year.

In most cases, where an object is to be coated with
lacquer, a base or substrate of the required material
(e.g. wood, bamboo or metal) is prepared and shaped. The
surface is sealed, often with the treated sap, used alone
or with a diluent such as ligroin. When this has hardened
the foundation is applied, usually layers of treated sap
mixed with rice starch paste, powdered fired clay or
stone dust. Cloth (frequently of hemp) may be applied to
strengthen edges and prevent warping. The treated sap is
scooped from the storage tub and mixed with a small
amount of water and clay dust or other filler, if
required, until the correct consistency is achieved. It
is then smoothed onto the surface, usually with a broad
wooden spatula or a flat hair brush, as shown in Figure
3, and allowed to harden. Each time a layer of lacquer is
applied the object is placed in a humid environment which
is kept as dust-free as possible, until the layer has
hardened. Fine craftsmen were reputed to take their
objects out on boats to keep the surface unblemished.
More usually, the object is placed for 1 - 3 days in a
wooden cabinet which is kept moist. Each layer is usually
rubbed down before the application of the next, using
abrasives of suitably increasing fineness; whetstone
powders, charcoal and finally calcined horn ash with an
oil as medium for the finest layers.

After a suitable foundation has been built up,
layers of treated sap alone or mixed with pigment may be
applied. Figure 4 shows a thin-section of a sample from
a Japanese helmet. Beneath a thin opaque layer of surface
dirt lie three upper layers of transparent black lacquer
and a thick band of foundation, containing clay minerals
and a large hole where some grains have become detached.
Underneath this is a thin layer of transparent lacquer
and below this the opaque steel base. Although the upper
layers appear black in reflected light, transmitted light
reveals their transparency (and the absence of any carbon
pigment) and shows the difference in hue between the
different layers.

A wide range of decorative techniques and finishes
determine the details of the application, especially in
the final layers, but basic rules apply to all. The
lacquer must be applied in relatively thin layers or it
will not harden throughout. The lacquer is rubbed down,
partly to ensure the smoothness and evenness of the
surface, but possibly also to ensure good bonding between

layers. The ambient temperature should not be too high or
too low, but the most critical factor of all appears to
be the humidity. If this is too low the lacquer will not
harden completely, at least 80% RH being generally
considered necessary for polymerising treated sap.

Lacquering is generally a slow process and major
items could take years or even generations to complete. A
variety of lacquers might be used on one item, ranging
from freshly treated saps for the foundation layers to
aged treated sap or treated sap from old trees for the
final surface.

Figure 3 Applying the foundation layers to a wooden box.

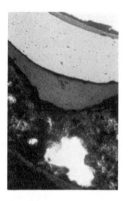

Figure 4 Thin-section of
black lacquer on a helmet.

3 ANALYSIS

The lacquer itself when hard is dense, smooth and
remarkably durable, seemingly resistant to water,
solvents, dilute acids and alkalis. Hot, concentrated
acids break it down and direct heat causes it to char to
a hard porous mass. Aged lacquers, although they may be
discoloured by solvents, show no tendency to dissolve. It
is not surprising then, that early scientists found this
material as baffling as it was beautiful. Lacquered
objects were exported to Europe in the mid-17th century,
during a flowering of scientific endeavour and the new
material aroused the interest of Robert Boyle - but a
series of analyses for the Royal Society for Improving
Natural Knowledge[6] were a failure. Its origin as a tree
sap was known and descriptions were published of its
extraction and use in China[7,8]. Its incomprehensibility
to the scientists of the period and the toxicity of the
sap, which can cause blistering of the skin, meant that
attempts by craftsmen in Europe to duplicate the effect
of lacquer were restricted to producing high gloss
solvent-based or oil-based varnishes.

It was not until the expansion of trade with the Far
East in the late 19th century and another great

acceleration in scientific endeavour that new methods of organic analysis began to reveal the first clues. Working from the soluble sap, rather than the insoluble hardened lacquer, Ishimatsu published one of the first analyses in this country in 1882 in the Memoirs of the Manchester Literary and Philosophical Society[9]. Since then much work has been carried out on the sap itself and on hardened lacquer, both ancient and modern.

Nowadays "urushi" (the Japanese name for the product used to produce lacquered articles) and "lacquer" are frequently used when referring to the sap, the treated sap before use and the hardened lacquer, leading to some confusion. In this article the raw product extracted from the tree will be referred to as "sap", the filtered, heated and mixed sap used to lacquer articles as "treated sap" and the finished product as "hardened lacquer". "Lacquer" has for so long been used to describe all three stages that it is hard to relegate it to just one part.

Analysis of the Sap

The first experiments relied on separating the sap into fractions according to solubility in alcohol or water, but it was not until the development of chromatography, and infra-red spectroscopy that the true complexity of the system began to be revealed. Although some analyses of saps from other species have been undertaken[10,11,12], most attention has been paid to the sap from R. verniciflua; this is the origin of much of the lacquerware dealt with by conservators.

It was soon realised that the sap was a water in oil emulsion with the principal polymerising element contained in the oil phase. Although briefly thought to be an acid[13], it was discovered to be a mixture of dihydric phenols or catechols, named urushiol[14] after the Japanese word "urushi". This group comprises on average about 65% of the sap, although the high water content of the fresh sap may decrease this to 50%. The water phase was found to contain "gummy substances", now identified as mono- oligo- and poly- saccharides in solution. A small amount of nitrogenous substances precipitated as insoluble material when the sap was separated with acetone are now known to be glycoproteins[15]. A number of metaloproteins have been isolated from the sap, including two enzymes, laccase and stellacyanin. The constituents of a typical sample are shown in Figure 5.

Majima sketched out the skeleton of urushiol, a dihydric phenol with a side chain of 15 carbon atoms and suggested that at least three similar compounds were also present, with one, two or three double bonds in the side chain[16]. Further studies revealed that some of the side chains can contain up to 17 carbon atoms, and that the double bonds may be in a variety of positions. The nucleus may also differ, both in the position of the side groups and in the number, at least two forms of a monohydric phenol having been detected[17]. The most common structures are shown in Figure 6. The most common is the molecule with three double bonds, two of them conjugated.

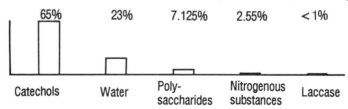

Figure 5   Average composition of sap of R. verniciflua

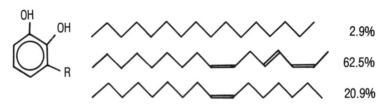

Figure 6 The original urushiol skeleton and its two most common variants.

The two other constituents of the lacquer to have received a similar amount of analytical interest, the enzymes laccase and stellacyanin, are probably distributed in both the water and oil phases of the emulsion. These are both copper metaloproteins; laccase containing 2 active and 2 non-active cupric atoms per molecule and stellacyanin containing one. Although the structure of stellacyanin has been investigated in some detail[18,19], its function in the polymerisation of lacquer is completely unknown. It contains a sulfhydryl group and three tryptophans, so it is possible that it exerts a regulatory function on other enzymatic action.

Laccase is strongly oxidative and takes an active and initiatory role in polymerisation. The amount present appears to influence the formation of the lacquer films and their subsequent properties. Too low a percentage of laccase appears to inhibit polymerisation, leading to the formation of incompletely hardened films. Too high a concentration may lead to overaccelerated polymerisation, resulting in films which have poor mechanical properties and which may be more susceptible to oxidation.

Among the saccharides identified from the sap are glucuronic acid[20], galactose, galacturonic acid, galacturonosylgalactose[21], galactopyranan, arabinose, rhamnose and various residues[22]. These have been shown to have branched structures and to form fibrous bundles[23], which may augment the durability of the lacquer film.

The other glycoproteins identified in the sap are little understood. Their presence may be important in preserving the emulsified state of the sap. Glycoproteins, proteins to which polysaccharides are linked, are valued for their mechanical properties, especially as stabilisers, but may also be useful as agents affecting reactions at boundaries.

4 POLYMERISATION

Polymerisation of the urushiol molecule has been studied for some time and the major linkages recognized. Recent improvements in analytical techniques have helped to elucidate further details, though the work is by no means complete. In particular, studies have indicated that there is a considerable difference in the properties of lacquer films formed by the polymerisation of the raw sap and those formed from the seasoned, treated sap. Films from the untreated sap tend to cross-link more, gradually increase in density and are more oxidisable.

The main difference between the raw sap and the treated sap originates in the process of stirring and gentle heating to drive off much of the moisture. Treated sap may have the water content reduced from 25-35% down to 12% or as little as 2%, influencing the rate of oxygen transfer and thus of oxidative enzyme action during polymerisation. Kumanotani has looked at this stirring treatment in detail and determined that the early stages of polymerisation take place then as well as water loss[23].

As the water content is reduced, the polysaccharides in the aqueous phase become more concentrated and are deposited. The physical action of the stirring breaks them up into small particles and distributes them throughout the oil phase in close association with the urushiol molecules. This may be assisted by some interaction between the urushiol and the glycoproteins. The stirring action may also help in the distribution of atmospheric oxygen throughout the sap, in a manner analogous to the preparation of drying oils for paint. The activity of the laccase appears to be considerably lessened in the treated sap and it is not yet clear if this is influenced by the heating, exposure to air or other reactions within the sap.

Molecular structure

The polymerisation of the urushiol molecule is initiated by the oxidative enzyme laccase, forming a semi-quinone radical (Figure 7). The Cu^{2+} atoms in the laccase molecule are reduced to Cu^+ and must be restored by oxidation. This process produces water, which aids the diffusion of oxygen through the less permeable polymerised surface.

The semi-quinone radicals form C-C bonds with each other or with the urushiol nucleus, forming diphenyl dimers, some of which may be enzymatically oxidised to form dibenzofuran compounds. The semi-quinone radicals may also react together to form an urushiol-quinone (Figure 8). The formation of the o-quinone and the dimers occurs within the first few hours of polymerisation after the lacquer film has been spread on the object. Although the o-quinone can be formed in sap films at 0%RH (presumably due to the high moisture content of the sap), the formation of the diphenyl dimer, vital to the hardening of the film, cannot take place unless there is further moisture present.

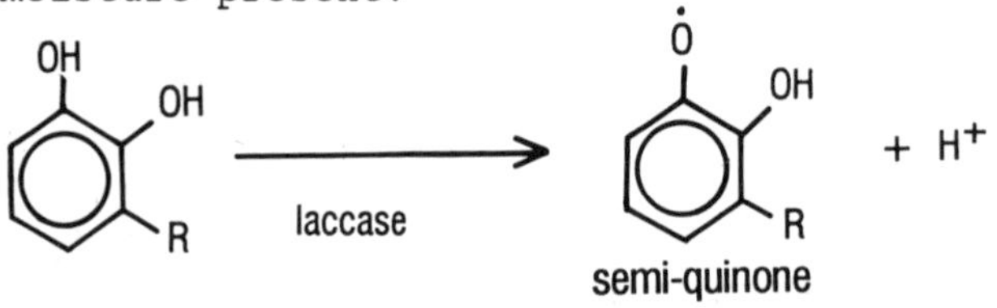

Figure 7 Initial polymerisation of the urushiol molecule.

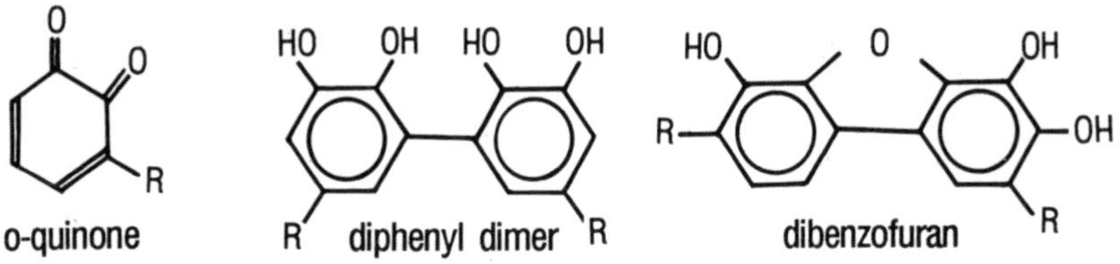

Figure 8 Further products.

Over the next few days the skeleton of the polymer is built as further bonds are formed between the unsaturated side-chains present and the nuclei of the urushiol molecules and their polymerised products (Figure 9).

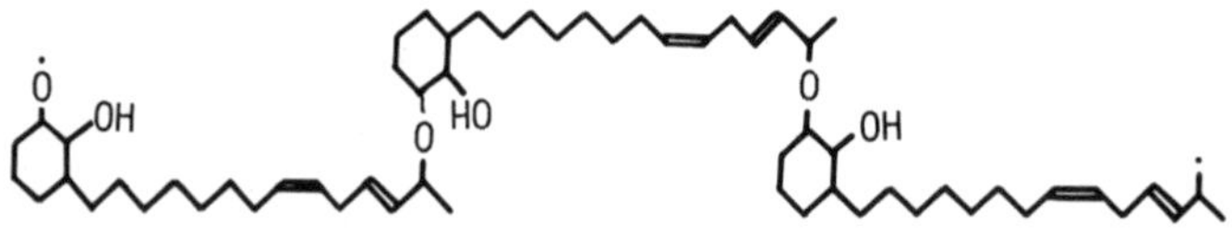

Figure 9 Simplified polymer skeleton

In treated sap, the formation of C-O bonds linking the side chain of one molecule to the nucleus of the next may be more common, leading to a more stable polymer. In the polymerisation of the untreated sap, the higher concentration of water favours the formation of C-C bonds, leading to more crosslinking of the chains.

Varying the relative humidity in which the sap is polymerising affects the process significantly. Below 43% RH polymerisation did not take place. At 95% RH the polymerisation was rapid and the film formed was rough and matte. The optimum RH was 55%, which produced a film which appeared glossy and smooth and matched the physical properties of films prepared from the treated sap[24]. Treated sap, which has a very low water content, requires a very high RH to polymerise successfully, generally above 80%. The process of reducing the water content of the sap and the practice of leaving the treated sap to season, during which time laccase activity diminishes enable the craftsman to regulate the speed of polymerisation by the addition of further water and the controlling of the ambient humidity. Slowing the rate of polymerisation produced more attractive and more stable lacquer films.

After the film has hardened for several days, it begins to increase in weight as atmospheric oxygen is absorbed. The unsaturated side chain may be oxidised, forming peroxides which link with the catechol nucleus or crosslink with other sidechains. This crosslinkage is believed to be more common in films from untreated sap and may resemble the formation of three-dimensional structures in drying oils.

Physical Structure

It has long been observed that the lacquer changes in appearance as it polymerises. From a greyish opaque liquid it gradually darkens and becomes translucent or even transparent, depending on the grade. A sample of sap was observed under the microscope as polymerisation took place under ambient conditions (approximately 20°C, 50% RH). The sap was found to be greyish in appearance under transmitted light, or a lighter cream colour in thinner areas. The fine globular nature of the emulsion was immediately apparent, and the presence of numerous small plant cells, crystals similar to oxalates and scraps of apparently polymerised lacquer was noted. Within 10 minutes the samples had begun to change slightly to a warmer, buff colour. 15 minutes after exposure to air the grains were noticeably larger. After 25 minutes the granulation appeared to diminish, in some areas to almost disappear. One hour after the first exposure to the air the granulation had reappeared and the colour was changing to orange. Twenty four hours after exposure the surface could be seen to be rough and granular, and appeared red to transmitted light.

Kumanotani has examined the heterogeneity of the lacquer film to see whether the lacquer polymerised evenly throughout or whether the hardened film on the surface was different in construction. He showed that the surface layer was richer in polysaccharides and in urushiol dimer units. The concentration of

polysaccharides may have been induced by the formation of
water below the hardened surface layer in the first stage
of polymerisation - a phenomenon studied by using ATRIR-
spectral analysis[25]. He etched the surface of a treated
sap film, revealing grains approximately 1000 Å in
diameter arranged in densely packed chains[26].

He postulates that each grain is a giant polymerised
urushiol molecule coated with a layer of polysaccharides
which are bonded to it by glycoprotein action. Although
the polysaccharides could absorb humidity, the bonding
between the glycoproteins, polysaccharides and the
hydrophobic urushiol polymer at the surface of the grains
may prevent easy ingress by water and thus by atmospheric
oxygen. Kumanotani believes this may help to explain the
greater stability of films made from treated sap. In
addition, the greater concentration of urushiol dimers
may help to prevent radical chain reactions spreading
into the body of the lacquer during deterioration.

In untreated sap, which has a higher water content
and where the globules containing the polysaccharides
have not been broken up and distributed by the stirring
treatment, he suggests the polysaccharides do not deposit
around the urushiol grains, but instead form islands in a
matrix of urushiol polymer, which is thus exposed to
atmospheric water and oxygen and more easily oxidised and
degraded. This may be supported by earlier studies which
revealed the presence of much coarser, crystalline grains
in sap films (approx 3 - 10 μ)[27].

What is not yet clear is how far the granular
urushiol structure extends into the lacquer layer. In
thicker layers, some polysaccharides can be precipitated,
leaving a urushiol-rich band sandwiched between the
surface and base. Since the normal process of lacquering
would require that the surface of each lacquer layer be
smoothed down and polished with abrasives, some of the
grains will be removed. Could this lead to the removal of
the entire polysaccharide-rich upper layer, or would
sufficient remain to protect the newly polished surface ?
Does the polishing affect the structure of the lacquer
film in any way and could it enhance or diminish the
stability of the layer ?

Additives

The presence of fillers, diluents or pigments may
also affect the polymerisation process. Red iron oxide
has been shown to accelerate the initial stage of
polymerisation and to retard the later stages[28].
Oxidising agents may speed the initial stages of
polymerisation too rapidly, leading to imperfect film
formation[29]. Drying oils retard the rate of hardening
and affect the physical properties of the film.
Temporary hydrogen bonds appear to form between the

hydroxyl group of the urushiol and the ketone group of the oil[30].

5 DETERIORATION

Deterioration of the lacquer must be separated from deterioration of the base which is often the primary cause of damage. Cracking, splitting, shrinkage, corrosion and degradation of the base may be accelerated after the lacquer layer is breached. For this reason, sampling from objects is best restricted to those where the lacquer layer shows some damage. In most cases, as the base is exposed to changes in humidity and temperature and degrades, the lacquer cracks and breaks away from the substrate.

Examples of early Chinese red and black lacquerware on wood[31] from the Zhou and Ch'in dynasties (11th - 8th c. and 3rd c. BC), some excavated from waterlogged conditions, exhibit a curious plasticity of the lacquer which clings to the substrate and deforms with it into grooves and wrinkles. It is possible that the lacquer used comes from a different species, but it seems more likely at present that the change in physical properties is brought about by the manipulations of the craftsmen of the period or the effect of waterlogging. More recent material excavated from waterlogged conditions does not show this plasticity, the lacquer retaining its shape even when the substrate is deformed or destroyed.

Lacquer films from the untreated sap can break down by oxidation, cross-linking and oxidative degradation[32]. The treated sap film may deteriorate by destruction of the binding between grains, causing the loss of the upper layer of grains as a loose powder and leading to the exposure of a fresh, rougher surface of grains[33], as well as by chain scission, leading to disruption within the grains. Ultra-violet light has been known for some time to be harmful, leading to the formation of paler surfaces, powdering and even fine cracking. The first two effects are probably the result of the intergranular disruption, though the cracking may be influenced by shrinkage as part of the skeleton is destroyed. Lacquers should be protected from light with a wavelength below 365nm. Other tests have shown some lacquers to be sensitive to non-UV light and heat[34]. In European climates the main hazard is often the extreme changes in relative humidity. Lacquer settles into an equilibrium with the ambient moisture and deteriorates if allowed to become too dry. Excess humidity may also contribute to deterioration, if absorbed by the hydrophilic polysaccharides and glycoproteins, by increasing the rate of oxygen diffusion within the film.

Kenjo states that part of the decomposed lacquer volatilises, presumably as chain scission or oxidation

splits off low molecular weight compounds. Diketone
compounds may also be formed by oxidative degradation[35].
She also produced photomicrographs of surfaces damaged by
exposure to U-V radiation, which produced small black
spots, often with a white border[36]. Similar results were
noted after exposure to high lux levels (above 200 lux)
of light containing no U-V.

Examples of red, black and gold lacquers exposed to
U-V light showed marked surface changes when examined
using a scanning electron microscope. The black lacquer,
which had been smooth, cracked into a series of flat
platelets (Figure 10).

 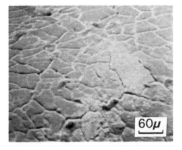

Figure 10 Black lacquer - before, after exposure to U-V

The red lacquer, which had a surface distorted by
the tiny grains of cinnabar, showed signs of deep cracks
running along the surface (Figure 11).

Figure 11 Red lacquer - before,   after exposure to U-V

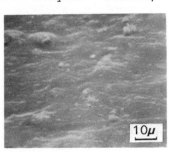 

Figure 12 Gold lacquer - before,   after exposure to U-V

The gold lacquer, which had a smooth surface overlying the minute particles of gold dust showed the most marked change, breaking up into rough lumps and masses with no particular orientation (Figure 12).

The presence of additives within the lacquer may accelerate deterioration, either by chemical interaction or even by the physical disruption caused by small structures (of pigment, filler or metallic decoration). Red lacquer pigmented with cinnabar (mercuric sulphide) appears to be less stable to light than that using cadmium sulphide[34]. Cinnabar is well known for accelerating the oxidative degradation of drying oils[37], and may have a similar effect in lacquer. Most red lacquers appear to be far less stable than black lacquer. This may be because the red pigments are actively destructive, because they form heterogeneous structures within the lacquer, or because the interaction between the lacquer and the iron compounds provides a greater stability. Lacquers with carbon powder incorporated may benefit from the energy absorbing protection that carbon exhibits in other polymers.

At present little work has been presented on the mechanisms of deterioration. As the structure of the polymer becomes more thoroughly understood, conservators and analysts alike will be able to concentrate more on the deterioration of this enigmatic material in an attempt both to understand it and to minimise it.

6 CONCLUSION

Lacquer is a fascinating material with a stability which can sometimes be frustrating to those seeking to analyse it. Studies on the unpolymerised material have enabled much of its complexity to be unravelled. Studies on the hardened material have examined modern lacquer for clues as to its formation and deterioration and ancient objects in an attempt to discover the details of their manufacture, origin and provenance. Observing the techniques used by craftsmen ancient and modern can give further clues. Increasingly it seems that the more the complex structure and processes within the material are unravelled, the more often we find that craftsmen centuries ago had recognised these properties and developed techniques to combat or incorporate them.

Some of the greatest advances in our knowledge of this material in the last thirty years have been the result of the work of Ju Kumanotani and Toshiko Kenjo in Japan. No work on the analysis of lacquer can fail to acknowledge the debt to them.

Acknowledgements
Jonathan Ashley-Smith and Nick Umney at the Victoria and Albert Museum, for their generous help with samples.

Birkbeck College, University of London, for SEM
photography.
Dept of Conservation and Materials Science, Institute of
Archaeology, University of London.

References

1   E. J. Kidder, 'Ancient Peoples and Places:
        Japan',Thames & Hudson, 1959.
2   Y. Kuraku, 'Urushi', N. S. Brommelle and P. Smith
        (eds.), Getty Conservation Institute, California,
        1988, 45.
3   Y. M. Du, 'Urushi', N. S. Brommelle and P. Smith
        (eds.), Getty Conservation Institute, California,
        1988, 194.
4   J. Hu, in 'Conservation and Restoration of Cultural
        Property: Conservation of Far Eastern Objects',
        Tokyo National Res. Inst. of Cultural Properties,
        Tokyo, 1980, 89-112.
5   also referred to as Rhus vernicifera
6   H. Huth, 'Lacquer of the West', University of
        Chicago, Chicago, 1971.
7   F. Bonnani, 'Vera vernix Sinica, Musaeum
        Kircherianum Romae', Rome, 1709.
8   Pere d'Incarville, Memoires de Mathematique et de
        Physique, presente a l'Academie Royale des Sciences,
        1760, III.
9   S. Ishimatsu, Memoirs of the Manchester Literary
        and Philosophical Society (3rd Series) 1882, 7, 249.
10   Y. Oda et al., Nippon Nogei Kagaku Kaishi, 1962, 36,
        (6) 527-531.
11   G. Betrand et al., Bull. Soc. Chim.,1939, 6, 1690.
12   Y. M. Du, ibid., 1988, 190.
13   H. Yoshida, J. Chem. Soc., 1883, 43, 472.
14   K?. Miyama, Kogyo Kagaku Zasshi, 1877, 10, 107.
15   J. Kumanotani, 'Urushi' N.S.Brommelle and P. Smith
        (eds.), Getty Conservation Institute, California
        1988,   244.
16   R. Majima, Berichte der deutschen chemischen
        Gesell-schaft 1922, 55B, 172.
17   Y. M.Du, ibid, 1988, 191.
18   L. Morpurgo et al., Biochim. Biophys. Acta, 1972,
        271 (2), 292-299.
19   C. Bergman et al., Biochem. Biophys. Res. Comm.,
        1977, 77 (3), 1052-1059.
20   Y. Oda et al., Nippon Nogei Kagaku Kaishi, 1963, 37
        (11), 663-667.
21   Y. Oda et al., Nippon Nogei Kagaku Kaishi, 1964, 38
        (2), 64-70.
22   Y. Oda and K. Koshiba, Agr. Biol. Chem. (Tokyo),
        1964, 28 (10), 678-685.
23   J. Kumanotani, ibid., 1988, 245.
24   T. Kenjo, 2nd International Symposium on the
        Conservation and Restoration of Cultural Properties,
        Tokyo, 1978,   151-163.

25   J. Kumanotani et al, <u>Proceedings 11 of the 12th</u>
      <u>International Conference on Organic Coatings,</u>
26   <u>Science and Technology</u>, Athens, 1986, 195.
27   J. Kumanotani, ibid., 1988, 248.
      J. Kumanotani, <u>2nd International Symposium on the</u>
      <u>Conservation and Restoration of Cultural Properties</u>,
28   Tokyo, 1978,   51-62.
29   T. Kenjo, <u>Shikizai Kyokaishi</u>, 1971, <u>44</u>, 470-474.
30   J.Winter, personal communication.
31   T. Kenjo, <u>Hozon Kagaku</u>, 1977, <u>16</u>, 12-16.
      B. Millam and H. Gillette, 'Urushi', N. S.
      Brommelle and P. Smith (eds), Getty Conservation
      Institute, California,   1988, 199.
      Also see: Lien Cosmetic Box, M.H. de Young Memorial
32   Museum, San Francisco, USA.
      J. Kumanotani, <u>American Chem. Soc., Division of</u>
      <u>Organic Coatings and Plastic Chemistry</u>, 1981, <u>45</u>,
33   643-648.
34   J. Kumanotani, ibid., 1988, 248.
      T. Araki and H. Sato, <u>Kobunkazai no Kagaku</u>, <u>23</u>, 1-
35   24.
      K. Toishi and T. Kenjo, <u>Shikizai Kyokaishi</u>, 1967,
36   <u>40</u>, 92-93.
      T. Kenjo, 'Urushi', N. S. Brommelle and P. Smith
      (eds.), Getty Conservation Institute, California,
37   1988, 162.
      J.S. Mills and R. White, 'The Organic Chemistry of
      Museum Objects', Butterworths, London, 1987.

Stability and Function of Coatings Used in Conservation

E. René de la Rie

NATIONAL GALLERY OF ART, SCIENTIFIC RESEARCH DEPARTMENT, WASHINGTON DC 20565, USA

1 INTRODUCTION

Coatings used today in the conservation of historic and artistic works are mostly clear coatings, that is, they are unpigmented or contain relatively little coloring material. Depending on the application, such coatings are also called finishes or varnishes. This paper addresses the stability and function of coatings in conservation. Only organic coatings soluble in organic solvents are considered.

Coatings are used on furniture, paintings, sculpture, glass and many other historic and artistic works. A wide variety of coating materials is used, including both natural and synthetic resins. Coatings can have a primarily aesthetic function or be applied for protection.

Most resins form transparent and colorless films, even those that are yellow when observed in lump form, such as many natural resins. Coatings may be required to bring about certain optical effects, such as color saturation, that are not realized equally well by different resins. In certain applications it may be desirable that a coating brings no change to the appearance of an object. This is usually a difficult requirement to meet. Many protective requirements are also difficult to meet with clear coatings as will be outlined below.

Another requirement not easily met is that of solubility. It presents a major difference when compared to most industrial coatings. Very stable coatings, for example, are produced for the automotive industry, but these are generally baked or cured in another way to provide a crosslinked structure. Industrial research is largely in the area of thermosetting resins and is, as such, of little value to the conservation field. Although sculptures and furniture often can withstand polar solvents, many paintings require that hydrocarbon solvents of low aromatic content, such as mineral spirits (white spirit), be used. Few polymeric resins exist that are soluble in mineral spirits.

Coatings should have sufficient flexibility so that they do not crack. In some cases abrasion resistance may be required. In conservation, non-polymeric resins are often used. Natural resins used in conservation are all of low molecular weight. Gum mastic and dammar resin contain substantial monomeric fractions; their polymeric fractions are small and of moderate molecular weight. Shellac consists largely of oligomers. Synthetic low-molecular-weight (LMW) resins, such as ketone resins and hydrocarbon resins, are also oligomers. These LMW resins are too brittle for applications where considerable flexibility and abrasion resistance are needed. For many paintings, however, LMW coatings are necessary to obtain the desired appearance (Section 3). How much flexibility is needed in a picture varnish has been the subject of debate,[1-3] although it appears that relatively brittle coatings can be used on many paintings. Extensive hydrogen bonding between polar oxidation products, however, causes the brittleness of natural and ketone resins to increase dramatically when they age. Also, MS2A (chemically reduced ketone resin; Section 5), a frequently used resin in picture varnishes, is very brittle, even when unaged, due to hydrogen bonding between abundant hydroxyl groups.

Coatings used in conservation should be chemically and physically stable so that their initial optical properties, mechanical properties and solubility are maintained over long periods of time. Coatings should preferably remain soluble in the same solvents in which they were initially soluble.

In the following, an overview is presented of the protective and aesthetic functions of coatings and of the factors affecting their stability. Some common coating materials are discussed specifically, although no attempt has been made to be exhaustive. The role of secondary ingredients in coatings is briefly discussed and some suggestions for future research are made.

2 PROTECTIVE FUNCTIONS OF COATINGS

Introduction

Coatings may protect objects against deposits and abrasion, and provide a surface that is cleaned more easily than an uncoated surface. Coatings may protect wooden and other surfaces from weathering. They may be applied as moisture barriers to inhibit corrosion of metals and glass or to reduce dimensional changes in

wood, ivory and other materials (caused by variations
in relative humidity). However, because ultraviolet
(UV) radiation, gases and vapors often pass easily
through these thin organic films, the protection
provided by clear coatings may be unsatisfactory in
many cases.

Ultraviolet Radiation

    Solar radiation occurs in the UV from about 290 nm
and up. Clear coatings are generally transparent in
this wavelength range and often well below 290 nm.

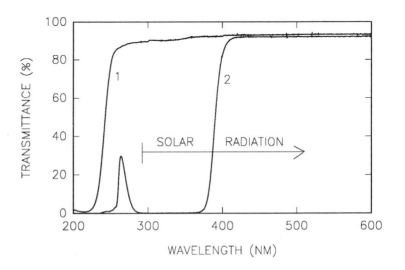

Figure 1 UV-vis transmission spectra of poly (iso-
butyl methacrylate) films of about 15 µm thickness with
(2) and without (1) a UV absorber.

    Only when a sufficient quantity of a UV absorber
(Section 4) is incorporated, can protection against
photochemical degradation (caused by UV light) be
achieved by a clear coating.[4,5] Figure 1 shows the UV-
vis transmission spectrum between 200 and 600 nm of a
poly (iso-butyl methacrylate) film and of one
containing 3% of the UV absorber Tinuvin 328 (a 2-
hydroxyphenyl benzotriazole derivative; Ciba-Geigy).
Use of UV absorbers in clear coatings is not common
practice in the conservation field. Their use is
highly recommended, however, in applications were UV
radiation is harmful to the object and cannot be
eliminated by other means, such as in outdoor exhibits.
(UV absorbers give little protection to clear coatings
themselves; Section 4.)

Table 1   Permeability coefficients[7]

Resin	O_2	H_2O
poly(vinyl alcohol)	0.04	79,000
poly(vinylidene chloride)	0.4	7.9
poly(methyl methacrylate)	67	470
poly(vinyl acetate)	220	4200
cellulose nitrate	390	3150
polyethylene (high density)	433	20

units for O_2: $[cm^3\ 100\mu m\ (m^2\ day\ atm)^{-1}]$ at 23 °C and 0% RH

units for H_2O: $[cm^3\ 10\mu m\ (m^2\ day\ atm)^{-1}]$ at 38 °C and 100% RH

Gases and Vapors

Besides being transparent to UV radiation, most clear coatings are also permeable to water vapor, oxygen and other vapors and gases*. An excellent review on the permeability of organic coatings has recently appeared[7] and the following is only a brief summary.

The transport of a gas or vapor through a polymer film depends both on its solubility in the polymer and its ability to diffuse through the film.[7,8] For simple gases and for water vapor through hydrophobic polymers, the permeability coefficient P equals the product of diffusivity D and solubility S: P=DxS. This holds true for systems in which D and S are independent of concentration and time. The permeability of polymers toward oxygen and other simple gases is largely controlled by diffusivity, while that of water vapor is largely determined by its solubility in the polymer.

Polar coating materials, such as poly(vinyl alcohol) (PVAL), are very good gas barriers, because of a high activation energy for diffusion (see Table 1; permeability data reported in the literature vary widely and are often determined under different conditions.[8]) Thomson has already shown that, due to its low oxygen permeability, a top coat of PVAL inhibits degradation of an underlying natural resin varnish.[1] It is important to realize that PVAL is one of the few materials that would have such an effect and that most coatings used by conservators are too permeable to oxygen to be able to inhibit autoxidative degradation of the layer underneath them.

---

*   Inorganic oxide coatings were recently proposed for glass protection, as these are much less permeable to water vapor than organic coatings[6].

PVAL is a poor barrier for water vapor, however, because of the solubility of water in the polymer (see Table 1). PVAL is generally known to crosslink upon aging, causing it to become insoluble,[1] although in a recent paper the claim is made that it degrades and remains soluble.[9] It adheres poorly to coatings soluble in mineral spirits[1] and it is therefore not a very suitable coating material.

Hydrophobic coating materials, such as waxes, on the other hand, are good barriers for water vapor because of low solubility of water in these non-polar materials (refer to the data for polyethylene in Table 1). They are poor gas barriers, however, because of high diffusion rates.

The best overall barrier properties are obtained with polymers of intermediate polarity, such as poly(vinylidene chloride) (PVDC) (see Table 1). Unfortunately, PVDC is a very unstable polymer and generally unsuitable as a coating material (Section 4). Other factors affect the permeability of coatings. Permeability decreases with increasing crystallinity and with increasing glass transition temperature. The low permeability of PVDC is caused in part by its high degree of crystallinity. Incorporation of plasticizers causes a lowering of the glass transition temperature and an increase in permeability. Polar polymers are plasticized by moisture and permeability may therefore vary dramatically with RH.

Coatings are often applied to metal surfaces to attempt to prevent corrosion. In many cases insufficient protection is obtained, as several reports in the conservation literature have indicated.[10,11] Corrosion inhibitors may be incorporated into coatings to improve their performance.[11]

How coatings prevent corrosion is still a matter of debate.[7] As the above discussion indicates, the permeability of most clear coatings is too high to remove oxygen and water from the corrosion process. However, a well-adhered coating inhibits the migration of ions across the surface and thereby corrosion.[7] Good adhesion of a coating to the metal substrate may be the single most important property controlling the inhibition of corrosion. The adhesive factor implies the necessity of using coatings with some polarity and may explain why epoxy resins perform relatively well.[10]

Coatings may protect silver from tarnishing. Tarnishing is caused by the reaction of silver with hydrogen sulfide:

$$2Ag + H_2S + 1/2O_2 \longrightarrow Ag_2S + H_2O$$

The permeability coefficient for hydrogen sulfide is somewhere between that of oxygen and water for many polymers.[8] Oxygen and water are also needed for tarnishing to proceed.[12,13] De Witte tested several coatings for their permeability to H_2S, water vapor and water. He concluded that poly(vinyl acetate) (PVAC) and cellulose nitrate are good barriers for H_2S and that acrylics are less satisfactory.[13]

3 AESTHETIC FUNCTION OF COATINGS

Clear coatings generally alter the appearance of surfaces considerably by increasing color saturation and gloss and by making colors appear darker. As such they may be integral parts of artistic or historic works, as is often the case with paintings and furniture. When coatings are applied for protection only, this change in appearance may be undesirable, but it is generally difficult to avoid. The change in appearance occurs due to a leveling of the surface (Figure 2) and because the coating has a higher refractive index (RI) than air.[14] Old master paintings often cannot be seen properly without a good varnish. Dark areas in such paintings particularly require a varnish that eliminates scattering of white light caused by microscopically rough surfaces. Old master paintings therefore require LMW coatings, as these level to a greater extent than polymeric coatings. A coating with a relatively high RI may also be required.[14]

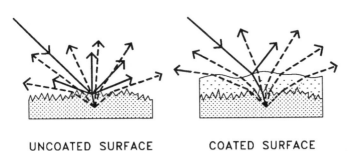

UNCOATED SURFACE   COATED SURFACE

——— Incident, reflected and scattered white light

- - - - - Colored light

Figure 2 Reduction of scattering of light at a colored and microscopically rough surface by application of a clear coating. As the amount of scattered white light decreases, color saturation increases.

Natural resins such as dammar resin, gum mastic and shellac have been used for centuries on paintings and furniture. Newly introduced synthetic LMW resins can produce appearances similar to those obtained with natural resins.[15,16] These are considerably more stable than natural resins (Section 5). The gloss of LMW coatings can be tailored by manipulating the surface of the coating during brush or spray application, by polishing the surface of the dried coating or by adding matting agents.[14,16]

4 STABILITY OF COATINGS

Autoxidation

Because of their large surface-to-volume ratio, coatings are subject to considerable exposure to the environment. Clear coatings, consisting of a thin film of organic material only, are particularly vulnerable to attack by oxygen in the air aided by other environmental factors such as UV radiation, humidity and pollutants.

Degradation of coatings occurs predominantly by autoxidation. Clear coatings are generally more susceptible to autoxidative degradation reactions than pigmented coatings, in which pigments may absorb UV radiation or stabilize the coating in other ways. In situations where UV light is present, photochemically initiated autoxidation is usually the main path of decay.

The initial reactions of the autoxidation process (see also the paper by McNeill in this publication), are the same for all organic materials, and follow a free radical chain mechanism, consisting of initiation, propagation and termination steps:[17,18]

$$\text{initiation:}$$
$$R\text{-}H + I^{\bullet} \longrightarrow R^{\bullet} + I\text{-}H$$

$$\text{propagation:}$$
$$R^{\bullet} + O_2 \longrightarrow ROO^{\bullet}$$
$$ROO^{\bullet} + R\text{-}H \longrightarrow ROOH + R^{\bullet}$$

$$\text{termination:}$$
$$2ROO^{\bullet} \longrightarrow ROOR + O_2$$

Alkyl radicals, $R^{\bullet}$, formed via hydrogen abstraction by a free radical initiator, $I^{\bullet}$, rapidly react with oxygen to form peroxy radicals, $ROO^{\bullet}$, which subsequently can abstract another hydrogen atom to form hydroperoxides, ROOH. Termination reactions are those between two free radicals, such as peroxy radicals, and give non radical products. Other free radicals that participate in the chain reactions are alkoxy radicals, RO•, and hydroxy radicals, HO•. These occur through scission of hydroperoxides, which readily decompose even in the absence of light. As hydroperoxides are also the primary products of autoxidation, the process is autocatalytic.

The chemical structure of a coating has a large influence on the rate of the autoxidation process. Certain functional groups promote hydrogen abstraction or homolytic bond cleavage leading to free radicals. These are primarily carbonyl groups, carbon-carbon double bonds (unsaturation), ether groups and tertiary carbon atoms. Hydrogens in α-positions to double bonds and those attached to tertiary carbon atoms are more easily abstracted than others:

Unsaturation:

$$\sim\!\!CH=CH-\overset{\overset{\displaystyle H}{|}}{C}H-CH_2\!\sim\ +\ I^{\bullet}\ \longrightarrow\ \sim\!\!CH=CH-\overset{\displaystyle \cdot}{C}H-CH_2\!\sim\ +\ IH$$

Tertiary Carbon Atoms:

$$\sim\!\!CH_2-\overset{\overset{\displaystyle H}{|}}{\underset{\underset{\displaystyle \sim\!CH_2}{|}}{C}}-CH_2\!\sim\ +\ I^{\bullet}\ \longrightarrow\ \sim\!\!CH_2-\overset{\overset{\displaystyle \cdot}{}}{\underset{\underset{\displaystyle \sim\!CH_2}{|}}{C}}-CH_2\!\sim\ +\ IH$$

Carbonyl groups undergo scission (Norrish type I or α-cleavage) in the presence of UV light and are therefore important sources of free radicals:

$$\underset{R_1}{\overset{\displaystyle O}{\overset{\displaystyle \|}{C}}}\underset{R_2}{}\ \overset{UV}{\longrightarrow}\ \underset{R_1}{\overset{\displaystyle O}{\overset{\displaystyle \|}{C^{\bullet}}}}\ +\ R_2^{\bullet}$$
$$\searrow\ R_1^{\bullet} + CO$$

Some coatings are inherently unstable and may degrade in part by non-oxidative mechanisms. PVDC, for example, loses HCl in an elimination reaction and cellulose nitrate decomposes hydrolytically while

releasing nitric acid. Both materials are among the
least stable coating materials in use in the
conservation field (see also section 5).

Stabilizing Additives

Stabilizing additives interfere with the
autoxidation process by removing free radicals (radical
scavenging), by absorbing UV light, quenching excited
states or by decomposing hydroperoxides.[4,18,19]
Stabilizers are widely used in industrial coatings but
are not commonly used in the conservation field.

Stabilizers should be selected with care as
negative side effects may occur. Many of the
antioxidants that are efficient radical scavengers
during thermal aging, such as hindered phenols, are
generally unsuitable for coatings, as they are not
light-stable and may develop colored transformation
products.[4,20] UV absorbers such as
hydroxybenzophenones and 2-hydroxyphenyl benzotriazoles
absorb UV radiation in the 300-400 nm range (Figure 1).
UV absorbers in clear coatings provide protection from
UV light to the surface underneath the coating. They
are of limited value for the stabilization of clear
coatings themselves because light penetrates the
coating to a considerable extent before it is
completely absorbed by the additive.[21]

Hindered amine light stabilizers (HALS) are light
stable radical scavengers and the most powerful light
stabilizers currently available. HALS are 2,2,6,6-
tetramethylpiperidine derivatives:

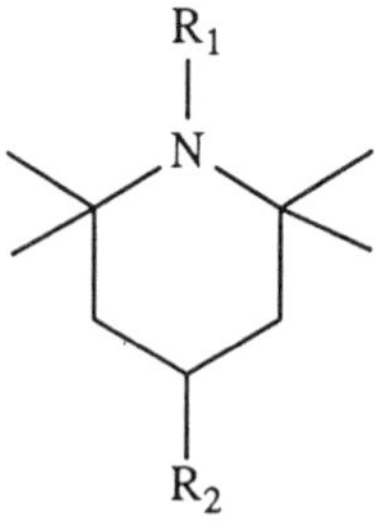

R_1 can be a hydrogen or an alkyl group. Peroxy
radicals oxidize secondary and tertiary hindered amines
to nitroxyl radicals, which are extraordinarily stable.
They do react with alkyl radicals, however, to produce
hindered aminoethers. Part of the effectiveness of
HALS stems from the fact that nitroxyl radicals can be
regenerated from hindered amino ethers by reaction with
peroxy radicals (this process is known as the Denisov
cycle):

$$\diagdown NH + ROO^{\bullet} \longrightarrow \diagdown NO^{\bullet} + ROH$$

$$\diagdown NO^{\bullet} + R^{\bullet} \longrightarrow \diagdown NOR$$

$$\diagdown NOR + ROO^{\bullet} \longrightarrow \diagdown NO^{\bullet} + ROOR$$

A variety of other reactions may occur.[18,19] Most HALS contain more than one group capable of scavenging free radicals in their molecular structure. In order to prevent loss by volatilization, HALS commonly are either monomers of relatively high molecular weight or polymers. HALS can be tremendously effective in coatings[15,21-23] and are the recommended stabilizers for coatings used in conservation.

Multiple Layers and Mixtures

Multi-coat systems or mixtures of resins have been used in the conservation field for a variety of reasons. Some reservation should be expressed with regard to the practice of applying a stable coating over an unstable one for the purpose of protecting the latter. Again, the easy passage of oxygen and UV radiation through the top layer will limit its protective function. About the only situation in which this would make sense is when a top coat containing a UV absorber is applied over a less stable coating on an object displayed in an environment containing UV light. Also the notion that an unstable resin would become more stable by mixing it with a more stable one is generally flawed. Only a stabilizing additive could play such a role. When applying different coatings on top of each other or when mixing resins, compatibility of the products should be a major concern.

5 COMMON COATING MATERIALS

Waxes

Waxes are common coating materials and are used either alone or mixed with other ingredients. Traditionally, waxes have been widely employed as moisture barriers, either to reduce dimensional changes in materials such as wood[24] and ivory[25] or to protect metal objects from corrosion[10,11]. Waxes, whether natural or synthetic, are relatively stable materials due to the absence of chemical groups that promote

autoxidation reactions. Beeswax and other natural
waxes consist of hydrocarbons, fatty acids and fatty
acid esters. Synthetic waxes often consist of
hydrocarbons only. Waxes have low abrasion resistance
and easily accumulate dust and dirt.

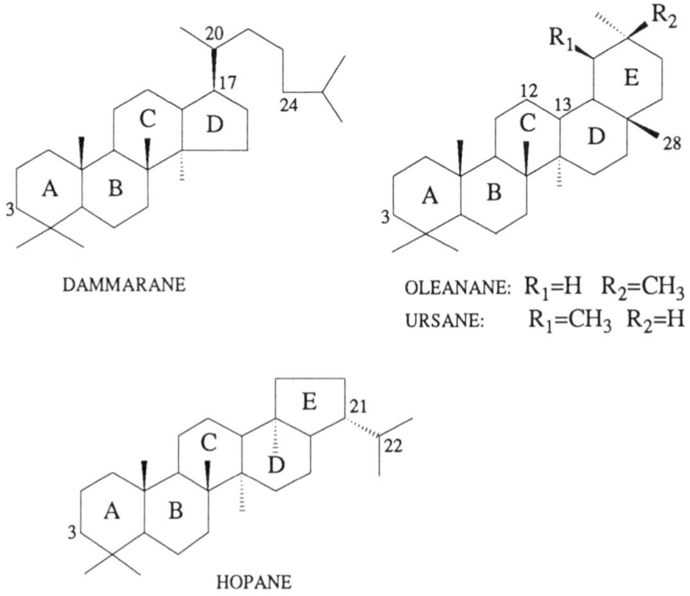

DAMMARANE

OLEANANE: R_1=H R_2=CH$_3$
URSANE: R_1=CH$_3$ R_2=H

HOPANE

Figure 3 Structures found in the triterpenoid fraction
of dammar resin. Functional groups susceptible to
autoxidation include ketone groups in the 3-positions,
double bonds in the 20 and 24 positions of dammaranes
and the 12-position of oleananes and ursanes, and
numerous tertiary carbon atoms. There is also a
polymeric fraction of moderate molecular weight and
largely unknown composition.[26,27]

Natural Resins

Natural resin coatings may be formulated as simple
solutions of the resins in organic solvents (spirit or
essential oil varnishes) or by boiling resins in a
drying oil such as linseed oil (oil varnishes). Both
have been used for centuries on furniture, sculpture
and paintings, although the use of oil varnishes has
declined dramatically.

Dried oil varnishes are insoluble because of
autoxidative crosslinking and are therefore difficult
to remove. They darken considerably with age. Spirit
varnishes containing dammar resin or gum mastic are

still commonly used on paintings. Shellac is still frequently used in furniture finishes. All natural resins are highly susceptible to autoxidative degradation due to the presence of chemical groups that promote these reactions. As mentioned before, these are primarily carbonyl groups, carbon-carbon double bonds, ether groups and tertiary carbon atoms. No attempt will be made here to describe the chemistry of natural resins as an excellent text on this subject exists[26]. Structures of triterpenoids occurring in dammar resin appear in Figure 3.

Natural resin coatings rapidly lose their initial properties. Autoxidation and subsequent degradation reactions lead to yellowing, embrittlement and formation of polar oxidation products, which cause loss of solubility in non-polar solvents (Figure 4).[28] Dammar films degrade primarily by photochemically initiated autoxidation followed by non-oxidative thermal reactions which produce yellow and fluorescent chromophores. These can be easily bleached by light. Upon degradation, the monomeric (triterpenoid) fraction disappears and oligomers are formed in dammar, but no insoluble crosslinked network is formed.[28] The degradation of natural resin varnishes poses a substantial problem in the field of painting conservation as deteriorated varnishes need to be removed frequently using polar solvent mixtures.[14] These solvents can be harmful to the paint layers as they may cause swelling and leaching of the binding medium.[2]

Recent work indicates that, in the absence of UV radiation below 400 nm, dammar resin films can be dramatically stabilized by adding HALS (Figure 4).[23,29] The use of HALS-stabilized dammar varnish is a viable alternative to using stable synthetic coatings for objects that are exhibited where UV radiation is eliminated.

Synthetic Coatings

Low-Molecular-Weight Resins. Synthetic LMW resins are used primarily in varnishes for paintings because they are capable of producing appearances similar to those obtained with natural resins.[14-16] Ketone resins have been used in the conservation field since the late 1940s. They have been available under trade names such as MS2, AW2 and Ketone Resin N. Currently, the BASF product Laropal K80 is available. Ketone resins, also known as polycyclohexanone resins, are condensation products of methyl cyclohexanones and/or cyclohexanone and are oligomers with number-average-molecular weights typically below 1,000.[30] In the conservation field ketone resins are generally used without additives although plasticizers are occasionally added.

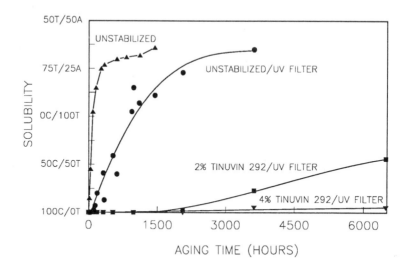

Figure 4 The effect of elimination of UV radiation and incorporation of the HALS Tinuvin 292 (Ciba-Geigy) on the solubility of dammar films. The films were aged in an Atlas xenon arc Weatherometer. A=acetone, C=cyclohexane, T=toluene; 50C/50T stands for a solvent mixture consisting of 50% cyclohexane and 50% toluene.

Ketone resins contain large numbers of ketone groups and hence are susceptible to Norrish type cleavage reactions. As a result of degradation, ketone resin coatings rapidly lose solubility in non-polar solvents, develop matte spots and embrittle.[30] They do not yellow as much as natural resin coatings. HALS are not capable of significantly inhibiting autoxidative degradation in Laropal K80.[15] Chemically reduced ketone resins, such as the currently available MS2A (Laporte), are considerably more stable than their parent products due to the elimination of ketone groups. They can be stabilized significantly with HALS.[15] Hydrogen bonding between the abundant hydroxy groups, however, makes such resins very brittle.

Recently introduced synthetic LMW resins, such as hydrogenated hydrocarbon resins and aldehyde resins, are much more stable than natural or ketone resins, although they are best stabilized with HALS.[15] Commercially available hydrogenated hydrocarbon resins are completely saturated cyclic hydrocarbon oligomers, with number-average-molecular weights between approximately 350 and 800 and are available from

several suppliers.[15] They dissolve in aliphatic
hydrocarbon solvents to give solutions with extremely
low viscosities. Aldehyde resins are condensation
products of urea and low molecular weight aldehydes.
One of these is particularly attractive as it combines
solubility in mineral spirits and stability.
Unfortunately, this product is currently not
commercially available.[15] These stable LMW resins are
beginning to be used in varnishes for paintings[16] but
could equally well be used in other applications in
conservation. Aside from their use as clear coating
materials they may also be useful as an inpainting
medium.

LMW coatings are more easily removed with solvents
than polymeric coatings. LMW resins form solutions of
low viscosity at high concentrations (typically up to
40 or 50% by weight) and can be removed with relatively
small amounts of solvent. Because solutions of
polymeric resins have manageable viscosities only at
low concentrations, large amounts of solvents must be
used to remove them. If volatile solvents are used, it
may be difficult to maintain the necessary amount of
solvent, resulting in the formation of a sticky mass.
In addition, autoxidative crosslinking is unlikely to
cause LMW resins to become insoluble, even if
autoxidation causes a major increase in molecular
weight.

<u>Polymeric Resins</u>. Synthetic polymers used in
coatings in the conservation field include PVAC and
several methacrylates. These are generally known for
their stability and have been investigated extensively
by Robert Feller and his co-workers.[2]

Nevertheless, autoxidative crosslinking reactions
cause some methacrylate coatings to become totally
insoluble upon aging.[2,9] Methacrylates prone to
crosslinking reactions are the higher alkyl
methacrylates, such as poly(n-butyl methacrylate) and
poly(i-butyl methacrylate), which are initially soluble
in mineral spirits (white spirit). The inhibition of
crosslinking in alkyl methacrylates by UV absorbers has
been the subject of several studies.[2,31,32] Feller
found that the induction time for the onset of
crosslinking in a copolymer of n-butyl and i-butyl
methacrylate increased by adding a hydroxyphenyl
benzotriazole UV absorber, either alone or together
with the antioxidant dilaurylthiodipropionate. A HALS
was also found to increase the induction time.[31] As UV
absorbers are relatively ineffective as stabilizers for
clear coatings (Section 4), the inhibition of the
crosslinking process by light-stable radical scavengers
such as HALS should be further investigated.

The lower alkyl methacrylates (which require solvents more polar than mineral spirits), such as poly (methyl methacrylate) and Acryloid (Paraloid) B72 (a copolymer of ethyl methacrylate and methyl acrylate[33]; Rohm and Haas), undergo chain scission rather than crosslinking when they age. Acryloid B72 is widely used in the conservation field as it is generally accepted as one of the most stable coating materials available.

PVACs of varying molecular weight are available. All have relatively low glass transition temperatures, which causes them to be rather soft and to pick up dirt and dust. PVACs of relatively low molecular weight, such as AYAB (no longer available) and Mowilith 20 (Hoechst), whose solutions have low viscosities, are used as inpainting media. PVACs have been reported to be very stable and chain break rather than crosslink.[34,35] Some studies on the photodegradation of PVAC have appeared, using low pressure and medium pressure mercury lamps.[36-39] As these lamps emit radiation at wavelengths below those found in sunlight, the results may not be relevant to the natural degradation of PVAC. Irradiation with a medium pressure mercury lamp causes PVAC to lose acetate side groups and the formation of conjugated polyenes (see also the paper by McNeill in this publication). Crosslinking and chain scission occur leading to the formation of an insoluble gel and a simultaneous decrease in molecular weight of the soluble fraction. Crosslinking appears to be significant only in high-molecular-weight PVACs.[36]

Some less stable synthetic resins continue to be used by conservators. Among these is cellulose nitrate, a semisynthetic polymer that has long been used in coatings for metals. Despite their inherent instability and the fact that many advise against their use in conservation, cellulose nitrate coatings remain immensely popular. Commercial solutions, usually in volatile esters or ketones, contain other resins, stabilizers and plasticizers. Examples of well preserved cellulose nitrate films of considerable age exist. Cellulose nitrate coatings may be relatively stable when not exposed to light.[40] Their popularity appears to be associated with short drying times, appearance and good adhesion to metal surfaces.[41] The primary mechanisms for cellulose nitrate decomposition have been reported to be acid catalyzed ester cleavage, homolytic scission of the nitrogen-oxygen bond and ring disintegration.[40] Hydrolytic decomposition is accompanied by the release of HNO_3 which subsequently catalyzes further hydrolysis.

PVDC or vinylidene/acrylonitrile copolymers are used as moisture barriers on wood, ivory and bone.[24,25,42] As PVDC has a lower permeability to water vapor than most other organic coating materials, it finds continued use despite its inherent instability.

6 SECONDARY INGREDIENTS

Coatings used in conservation often have a simple composition, that is they may consist of a single resin dissolved in a one-component solvent. Coatings in industry on the other hand may contain a number of ingredients, each with their own function. Ketone resins were developed as gloss modifiers for coatings and hydrocarbon resins as tackifying resins for adhesives. These resins are used industrially as additives and never as the sole solid ingredient of a coating. Similarly, synthetic polymers, such as PVAC and methacrylates, are always used together with other ingredients in industrial coatings. Solvent systems may also be quite complex. Nevertheless, many commercially available picture varnishes, for example, are simple solutions of ketone resins in mineral spirits.

Although a frequently used argument for using coatings of simple composition is to minimize possible side effects of secondary ingredients, such as their migration and exudation, or their adverse effects on stability, it may be difficult to obtain all required properties with a single ingredient. Stability, adhesion, flexibility and other properties of coatings currently used by conservators could be improved markedly by using additives. Obviously, low-molecular-weight plasticizers which may migrate or exude are not acceptable for most applications in conservation. Additives could be polymeric themselves, however, so that migration and loss due to volatilization are unlikely. Such additives should be carefully selected after substantial testing so as to make sure that they are compatible with the substrate and do not migrate or exude from the coating.

Stabilizing additives, such as HALS, currently rarely utilized in the conservation field, are highly recommended for coatings (section 4). Many different HALS are available that are compatible with a wide variety of substrates.[4]

Experiments to investigate the possibilities of increasing the flexibility of LMW coatings by using polymeric additives are underway at the National Gallery of Art in Washington. A LMW aldehyde resin obtained from BASF is compatible with poly(i-butyl methacrylate) and poly(n-butyl methacrylate). Simple removability tests indicate that at a 9:1 LMW resin/polymer ratio no insolubility due to crosslinking occurs in unstabilized films of this mixture during long term aging in a xenon arc Weatherometer. Hydrogenated hydrocarbon resins appear to be compatible with Kraton rubbers (Shell) and coatings with considerably higher flexibility may be formulated by incorporating these rubbers. The results of this work will be published.

7 CONCLUSIONS

A limited number of stable coating materials is currently available to conservators, including PVAC, the lower alkyl methacrylates, HALS-stabilized dammar (in the absence of UV light only) and some recently introduced synthetic LMW resins. Clear coatings may be integral aesthetic components of objects or may be applied to protect objects. Most organic coatings provide limited protection because they are transparent to UV radiation and because gases and vapors easily pass through them. The beneficial effects of secondary ingredients on phenomena such as stability, working properties and flexibility, should be investigated. The study of the adhesion of coatings and compatibility of different coating materials is also of great importance to the conservation field.

REFERENCES

1   G. Thomson, in 'Recent Advances in Conservation', G. Thomson, ed., Butterworths, London, 1963, p. 176.

2   R. L. Feller, N. Stolow and E. H. Jones, 'On Picture Varnishes and their Solvents', National Gallery of Art, Washington, DC, revised ed., 1985.

3   E. R. de la Rie, in 'Preprints to 8th Triennial Meeting of ICOM Committee for Conservation', International Council of Museums, Paris, 1987, p. 791.

4   E. R. de la Rie, Stud. Conserv., 1988, 33, 9.

5   J. Bourdeau, in 'Cleaning, Retouching and Coatings', J. S. Mills and P. Smith, eds., International Institute for Conservation of Historic and Artistic Works, London, 1990, p. 165.

6   M. W. Colby, T. J. Yuen and J. D. Mackenzie, in 'Materials Issues in Art and Archaeology', E. V. Sayre, P. B. Vandiver, J. Druzik and C. Stevenson, eds., Materials Research Society, Pittsburgh, PA, 1988, p. 305.

7   N. L. Thomas, Progr. Org. Coatings, 1991, 19, 101.

8   S. Pauly, in 'Polymer Handbook', J. Brandrup and E.
     H. Immergut, eds., John Wiley, New York, 1989, 3rd
     ed., p. VI/435.

9   J. Ciabach, in 'Resins in Conservation', J. O.
     Tate, N.H. Tennent and J. H. Townsend, eds.,
     Scottish Society for Conservation and Restoration,
     Edinburgh, 1983, p. 5-1.

10   S. Keene, in 'Adhesives and Consolidants', N. S.
     Brommelle, E. Pye, P. Smith and G. Thomson, eds.,
     International Institute for Conservation of
     Historic and Artistic Works, London, 1984, p. 104.

11   R. A. Turisheva, in 'Preprints to 7th Triennial
     Meeting of ICOM Committee for Conservation',
     International Council of Museums, Paris, 1984, p.
     84/22/41.

12   T. Stambolov, Stud. Conserv., 1966, 11, 37.

13   E. de Witte, in 'Bulletin Institut Royal du
     Patrimoine Artistique', Brussels, 1973/74, Vol. 14,
     p. 140.

14   E. R. de la Rie, Stud. Conserv., 1987, 32, 1.

15   E. R. de la Rie and C. W. McGlinchey, in 'Cleaning,
     Retouching and Coatings', J. S. Mills and P. Smith,
     eds., International Institute for Conservation of
     Historic and Artistic Works, London, 1990, p. 168.

16   M. Leonard, in 'Cleaning, Retouching and Coatings',
     J. S. Mills and P. Smith, eds., International
     Institute for Conservation of Historic and Artistic
     Works, London, 1990, p. 174.

17   G. Scott, 'Atmospheric Oxidation and Antioxidants',
     Elsevier, Amsterdam, 1965.

18   J. Pospisil and P. P. Klemchuk, eds., 'Oxidation
     Inhibition in Organic Materials', CRC Press, Boca
     Raton, FL, 1990, Vols. 1 & 2.

19   P. P. Klemchuk, ed., 'Polymer Stabilization and
     Degradation', American Chemical Society,
     Washington, DC, 1985.

20   E. R. de la Rie, Stud. Conserv., 1988, 33, 109.

21   P. J. Schirmann and M. Dexter, in 'Handbook of
     Coating Additives', L. J. Calbo, ed., Marcel
     Dekker, New York, 1987, p. 225.

22 B. Felder, R. Schumacher and F. Sitek, in 'Photodegradation and Photostabilization of Coatings', S. P. Pappas and F. H. Winslow, eds., American Chemical Society, Washington, DC, 1981, p. 65.

23 E. R. de la Rie and C. W. McGlinchey, in 'Cleaning, Retouching and Coatings', J. S. Mills and P. Smith, eds., International Institute for Conservation of Historic and Artistic Works, London, 1990, p. 160.

24 J. A. Brewer, Stud. Conserv., , 1991, 36, 9.

25 R. H. Lafontaine and P. A. Wood, Stud. Conserv., 1982, 27, 109.

26 J. S. Mills and R. White, 'The Organic Chemistry of Museum Objects', Butterworths, London, 1987.

27 E. R. de la Rie, Doctoral Dissertation, University of Amsterdam, 1988, Chapter 3.

28 E. R. de la Rie, Stud. Conserv., 1988, 33, 53.

29 E. R. de la Rie and C. W. McGlinchey, Stud. Conserv., 1989, 34, 137.

30 E. R. de la Rie and A. M. Shedrinsky, Stud. Conserv., 1989, 34, 9.

31 R. L. Feller, M. Curran and C. Balie, in 'Photodegradation and Photostabilization of Coatings', S.P. Pappas and F. H. Winslow, eds., American Chemical Society, Washington, DC, 1981, p. 184.

32 G. M. Nelson, and Z. W. Wicks, in 'Resins in Conservation', J. O. Tate, N.H. Tennent and J. H. Townsend, eds., Scottish Society for Conservation and Restoration, Edinburgh, 1983, p. 4-1.

33 E. de Witte, M. Goessens-Landrie, E. J. Goethals and R. Simonds, in 'Preprints to 5th Triennial Meeting of ICOM Committee for Conservation', International Council of Museums, Paris, 1978, p. 78/16/3.

34 R. L. Feller, in 'Preprints to 4th Triennial Meeting of ICOM Committee for Conservation', International Council of Museums, Paris, 1975, p. 75/22/4.

35 R. L. Feller, in 'Preservation of Paper and Textiles of Historic and Artistic Value', J. C. Williams, ed., American Chemical Society, Washington, DC, 1977, p. 314.

36 E. Y. L. Vaidergorin, M. E. R. Marcondes and V. G. Toscano, Polym. Degradation Stab., 1987, 18, 329.

37 K. J. Buchanan and W. J. McGill, Eur. Polym. J., 1980, 16, 309, 313 and 319.

38 C. David, M. Borsu and G. Geuskens, Eur. Polym. J., 1970, 6, 959.

39 G. Geuskens, M. Borsu and C. David, Eur. Polym. J., 1972, 8, 883 and 1347.

40 C. Selwitz, 'Cellulose Nitrate in Conservation', Getty Conservation Institute, Marina del Rey, CA, 1988.

41 D. B. Heller, in 'Preprints to 9th Annual Meeting', American Institute for Conservation of Historic and Artistic Works, Washington, DC, 1981, p. 57.

42 M. Bunn, in 'Archaeological Bone, Antler and Ivory', K. Starling and D. Watkinson, eds., United Kingdom Institute of Conservation, London, 1987, p. 28.

ACKNOWLEDGEMENTS

The author is grateful to Barbara Berrie, Sarah Bertalan, Christopher Maines, Albert Marshall and Mervin Richard for their help during the preparation of this paper.

Textile Conservation

J. S. Crighton

DEPARTMENT OF TEXTILE INDUSTRIES, UNIVERSITY OF LEEDS, LEEDS LS2 9JT, UK

The fibrous character of textile raw materials requires the presence of linear polymers of at least 10,000 molecular weight orientated in the long direction of the material, and with interchainic interactions provided by the existence of ordered volumes or by specific chemical interactions. When these features change then properties including fibrousness will also change and may be lost. These considerations apply to all fibres both manmade and natural. With the latter histological features as well as the molecular morphology determine the ultimate performance.

The museum curator and conservator are concerned with the prevention of degradation. In the current context "degradation" describes any change of behaviours away from those features expected of/associated with (a) changes in visual appearance, and/or (b) loss of those mechanical properties associated with the specific textile. The origins of changed behaviours may be at the structural and/or molecular levels. The energy sources to "drive" these degradative processes could be either (i) photolytic, from actinic radiation either by direct absorption, or sensitisation via "chromophores" contained within the total composition, (ii) mechanical, by stress transmitted into and through the structure, (iii) thermal or chemical, involving homolytic scission, hydrolysis or oxidation, or (iv) biological, including microbes which eat before digestion, or by bacteria which generate direct enzymic breakdown.

The objectives of the conservator can be recognised as follows:-
(A) The recognition of the occurrence of "degradation". The need is to achieve such at the earliest possible stage. The earlier that degradation is identified, the more effective are conservation methods likely to be in the sustaining of the original.
(B) The identification of the key contributors (degradation routes) leading to fibre breakdown.
(C) Initiation of action to prevent if possible, or at

least retard, the identified degrading pathways.
(D)   The conservation in a stable form of the degraded
textile.
Initial consideration will be given to (A), followed by
an examination of (B) and (C) for specific textile
fibres.   Finally, an overview of possible procedures to
achieve (D) will be presented.

Recognition/estimation of damage

     To achieve early recognition of change requires an
understanding of all the possible origins of actions
which   could   lead   to   degradation,   followed   by   the
establishment   of   the   most   sensitive   and   practicable
methods for monitoring with individual textile items.
One of the useful features is to recognise whether
degradation is homogeneous or whether it is located at
specific sites/volumes.   From such an assessment an
insight as to the potential sources of degradation can be
made.   It is desirable that all potential methods able to
recognise the occurrence of degradation be considered.
These should be critically compared, in the context of
their ability to identify the occurrence of specific
types of degradation.   Methods must be selected which
appear to be both practical and the most sensitive for
monitoring the presence of the critical degradation
pathways for the particular fibre.

     With fibrous polymer structures changed properties
may arise as a result of (i) chain breakage (either an
unzipping depolymerisation or a random scission), (ii)
chain crosslinking, (iii) the loss of, or chemical
modification to pendant side groups of the molecular
chain, (iv) the loss or creation of new interactions
between the side chains, as compared with the original,
or (v) structural changes, which could be either of an
histological or morphological origin.   Generally, with
macromolecules random scission is the most common cause
of property degradation, along with chemical changes
(loss or modification) of the side groups.   The most
popular methods used for estimating and monitoring damage
are listed in table 1.

Table 1   Polymer Structures: Methods for estimating
damage.

     Mechanical behaviour,
     Functional   group   characterisation   (chemical,
     physical),
     Structural characterisation,
     Solubility behaviour,
     Molecular weight assessment.

     Mechanical properties are frequently a demonstrable
indicator that damage has occurred.   However, depending
upon both the nature and the sites of the key degradation
processes, changed mechanical behaviours may only be a

secondary consequence of more radical change. Early
stages of change may not "show up" positively in the
observed mechanical properties.   The potential for
mechanically testing valuable artefacts is not a viable
option. Variations in mechanical properties can often be
directly correlated with structural changes in the
substrate.   Alternative procedures to assess these
structural changes do exist. Changes in the histology of
natural fibres can be assessed by microscopy.
Morphological changes associated with either the total
fraction of order, or the relative sizes of the ordered
volumes and their orientation can be identified by the
use of physical methods such as X-ray diffraction,
differential scanning calorimetry, differential thermal
analysis, density or sometimes infra-red spectroscopy.
Some of these methods can generate positive data from
extremely small samples.

    The functional groups present, and the loss or
generation of these during storage or display, can be
monitored by either physical or chemical methods,
depending upon the chemical nature of the group. Infrared methods, particularly as ATR and FTIR have potential
for application without additional damage to the examined
textile. With ATR, only the surface volumes, the most
likely site for change, are examined. Chemical methods
do require the "consumption" of small samples, and must
be of high sensitivity to be effective as adequate
monitors of chemical change.   As a result of chain
scissions or crosslink formation the solubility
characteristics of the polymer substrate will change.
Solubility behaviour is used to monitor damage, however
the choice of solvent is critical while the precise
locations of bond formation or scission, and side group
changes affect the potential sensitivity of this
technique.

    Molecular weight estimation is the most commonly
practised monitoring method. Ideally the molecular size
distribution should be assessed using size exclusion or
gel permeation chromatography. Even with this technique,
which assesses the effective volume occupied by the
dissolved macromolecules, distortions can arise in the
apparent molecular weight distributions, if changes in
solute shape occur. The need to employ the smallest
possible sample size has often meant that viscosity or
fluidity measurements and changes therein have formed the
prime monitoring procedure.   The most practical method
employs a suspended level dilution viscometer, with the
measurements of flow times for a fixed volume of
solutions of differing concentrations translated to yield
the intrinsic viscosity. This value is relatable to the
size of the polymer solute. Because there is normally a
range of sizes the viscometric technique only provides an
"average" molecular weight. The viscosity average
molecular weight is, for most polymer systems, close to
the weight average molecular weight.

In the early stages of those degradations which involve chain scission, independent of whether these scissions occur randomly or at specific sites along the chain, the most sensitive identifier of the occurrence of scission is ($\overline{M}_n$), the number average molecular weight or the number average degree of polymerisation ($\overline{dp}_n$) rather than the weight average ($\overline{M}_w$, $\overline{dp}_w$) or viscosity average values. (see Table 2)

Table 2. Targeted chain scission (at chain centre), with only one scission per initial molecule of an assumed monodisperse cotton cellulose of dp_n = 10,000. Influence on degree of polymerisation.

Percent of total chains degraded	$\overline{dp}_n$	$\overline{dp}_w$
0 (initial)	10,000	10,000
0.1	9,990	9,995
0.2	9,980	9,990
0.5	9,950	9,975
1.0	9,900	9,950
2.0	9,800	9,900
10	9,090	9,520
20	8,370	9,000

The number average molecular weight is achieved by indirectly "counting" molecules (e.g. osmotic pressure observations), or for the chemical assessment of the chain ends. Viscometric measurements are not the most sensitive observations to make, and efforts should be directed towards assessing number average values. Additionally, if side chain changes occur concurrent with chain scissions, the relationship between viscosity and the molecular weight is likely to alter; as a result, a further distortion in the observations will arise and any attempt at their interpretation should be made with caution.

Accelerated Methods

To recognise the most significant degradation mechanism relevant for a specific textile substrate, laboratory simulations can provide a helpful guide. These methods can also act as a useful testbed for the assessment of both degradation inhibition/retardation treatments and conservation procedures.

Potential simulations utilise accelerated ageing, in which the observations are interpolated to suggest performance levels for the treated textile on display or in storage. Great care and thought should be given to ensure the choice of those simulation conditions which accelerate only the specific degradation paths that dominate and determine degradation behaviours in the storage or display environment.

Accelerated tests are normally achieved by either

(i) the use of elevated temperatures, or (ii) the employ-
ment of stronger light intensities. Both options can
result in distorted observations and interpolations.
Their use must be practised with caution. Often, more
than one "degradation" route may be possible within the
storage or display environment, although only one route
is dominant. If there is more than one "degrading" pathway then, for each, the activation energies of their rate
determining steps are almost certainly different. The
consequences are illustrated in figure 1 for a system
where two possible routes exist. Their different
activation energies are reflected in the gradients of the
log (rate) versus reciprocal temperature plots for routes
A and B. If T_2 is the storage or display temperature,
then route B clearly dominates. However, if observations
are made at a higher temperature e.g. T_1, then the
degradation mechanism reflected by route A will dominate.
Observations based purely at T_1 cannot be interpolated to
T_2. Even observations at a range of temperatures will be
difficult to interpolate back to T_2. If, in raising the
temperature from T_2 to T_1, one passes through a second
order transition (the onset of segmental motion) or a
higher order transition (eg. the occurrence of side chain
movements) the character/activation energy of the rate
determining process can change significantly. As a
result a linear extrapolation back to T_2 will again not
be possible. Further complications can also result from
the raising of the temperature. There is likely to be an
accelerated loss of additives (such as plasticisers and
stablisers) which will certainly cause distortion of
observations and inhibit effective interpolation to the
"use" conditions.

When photodegradative routes provide the dominant element
in change, the common practise is to subject the test
samples to enhanced (by one order) illumination. The
characteristics of UK sunlight compared with a commonly
used Xenon arc are shown in Table 3.

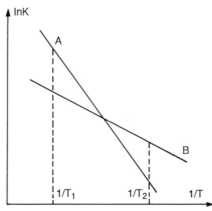

Figure 1. Activation energy plots for a system with two
concurrent reactions.

Table 3. A comparison of illuminations from sunlight and a Xenon arc (with and without filters).

Light Source	Power
UK (mid summer, 12 noon):	$1.0W/m^2/hr$
Xenon arc 1200 (filter):	$3.6W/m^2/hr$
Xenon arc 1200 (no filter):	$21W/m^2/hr$
Xenon arc 1200 (through filter and glass):	$0.2W/m^2/hr$

Care is normally taken to ensure that the spectral quality of the test illumination compares with that in the display environment, but the actual intensity of the incident radiation may cause distortions in observations away from those 'normally' found. At higher incident intensities there is a higher rate of formation of the primary active species. Higher concentrations of these species will therefore prevail than arise in normal use. Alternative or additional reactions can arise or assume a greater importance than normal. Quenching of activated species will also not be as effective. The rate of diffusive "escape" of the active species can be insufficient either to prevent their effective loss or, alternatively, their mutual interaction can result in a reduced apparent activity.

Degradation Mechanisms

To achieve efficient retardation or effective stabilisation from degradation it is necessary to identify and appreciate the primary routes through which property changes arise. Several of these will be considered here in the general context while others will be discussed under the consideration of individual fibres.

The effects of stress. Consider the potential energy diagram associated with a chemical bond in conjunction with its translation to the activated transition state for chemical reaction (Figure 2).

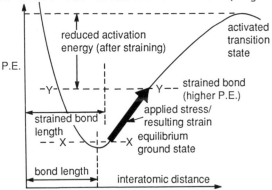

Figure 2. Energy diagram of an unstrained and strained chemical bond.

X-X reflects the equilibrium ground state of a particular
chemical bond.   The equilibrium bond energy and bond
length can be assessed from this position.   When this
chemical bond is strained, the result will be an extended
interatomic bond distance in which extra (strain) energy
is stored in that bond as represented by Y-Y.   The energy
required to achieve either bond fission or the activated
transition state for a chemical reaction involving that
bond is reduced, a consequence of straining the bond.
Textiles and their component fibres either in processing
or in subsequent use experience mechanical stresses.   The
prevailing stress levels will be propagated through the
structure and translated to strain which, within the
morphology of the fibre elements will be transmitted and
localised at specific chain segments, rather than spread
over the whole structure.   The amorphous chain segments
it is suggested may carry upto 20x the average stress
associated with the whole structure.   The result of such
localised stress variations within the whole will be the
effective   "activation"   of   bonds   towards   chemical
reaction.   Structures under stress and the specific
strained bonds which result are likely to undergo
accelerated chemical reaction with subsequent possible
degradation of the total structure within which they are
located.

    Within   those   polycrystalline   macromolecular
structures   which   possess   fibrous   character,   the
"organisation" of the molecular chains is likely to be
either of the fringed fibril (Fig. 3A) or folded chain
lamella (Fig. 3B).

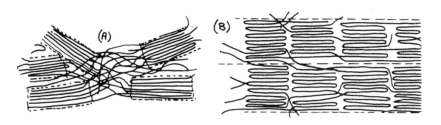

Figure 3.   A representation of ordered volumes in fibrous
structures.

Between the orientated ordered elements of the fringed
fibril chain segments there will be in the less organised
volumes segments some of which may be inherently strained
but definitely with some parts which are strained when
the total fibre structure is subjected to a tensile
stress.   A fibre element in a knitted textile can be
assumed to be strained into the shape

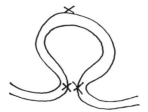

Figure 4. A representation of a filament or yarn loop in a knitted textile.

shown in Fig. 4. The "outer volumes of the loop, at the points X, will be subject to a tensile extension. In the folded chain lamella structures created by unfolding, slippage and twisting of ordered blocks in fibre drawing those bonds at the fold are positively strained and will thus be activated towards potential chemical reaction. In both the Fig. 3 morphological representations the "activated" elements are accessible. Evidence for accelerated degradation when stressed has been reported for polypropylene,[1] high density polyethylene,[2] cotton cellulose[3] and for rayons.[4]

Photodegradation. Sunlight at the earth's surface has, in general, an approximate composition of 5% UV, 40% visible and 55% Infra red. The quanta of highest energy are those of the ultra violet. The sensitivity of a fibre to photo induced change will depend upon the individual bond strengths and the sizes of the energy quanta absorbed, (ie. which wavelengths), and thus upon the total chemical composition of the fibre, to include not just its macromolecular composition but all additives (including dyes) as well as impurities. Decomposition or oxidation may result as a consequence of the energy absorption; either may produce a "loss" of appearance, changes in mechanical properties or in other "desirable" features. One can note that significant absorption in the UV arises when aromatic structures are present, e.g. in polyphenylene oxide, which darkens when exposed to illumination of less than 400nm. This characteristic can be utilised to provide a practical calibrant of the irradiation source. An approximate relative rating of sensitivity to sunlight is illustrated by:-

(LEAST   Silk ⟩   ⟩   ⟩   ⟩Acetate (MOST
STABLE) Nylon⟩ cottons⟩ rayons⟩ polyester⟩ Acrylic STABLE)

A demonstration of the influence of photodegradation on fibre strength is reflected in Fig. 5. The clear periodicity in behaviour demonstrates the photolytic component.

A more revealing comparison of susceptibilities to light is demonstrated in Table 4.

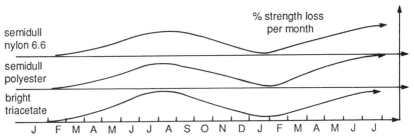

Figure 5. Relative monthly losses of strength following outside exposure in UK.

Table 4 Relative fibre strength losses when exposed to sunlight.

Behind Glass	Without Glass
Bright Orlon (Acrylic)	Bright Acrylic
Semidull Orlon (Acrylic)	Semidull Acrylic
Bright Dacron (Polyester)	Bright Polyester, bright Acetate
Semidull Dacron (Polyester)	Bright polyamide, rayons, cottons
Bright Acetate, bright polyamide rayons, cottons	Semidull polyester
Silk, semidull fibres (except Orlon, Dacron)	Silk, semidull fibres (except Orlon, Dacron)
Most matt fibres	Most matt fibres

DECREASING RESISTANCE

It is immediately apparent that the semidull and matt fibres are more sensitive to photodegradation than their otherwise equivalent bright fibres. Manmade fibres as directly fabricated from the viscous macromolecular solution or melt are inherently bright. For many applications this characteristic is not desired, and the common practice is to add to the solution or melt, and thus to the resulting fibre, a delustrant which is usually titanium dioxide. It is this titanium dioxide which sensitises the base structure to photodegradation (see ref. 5). Property losses are accelerated in air and with increased moisture levels. Allen et al., have suggested that the mode of action of the titanium dioxide is via the generation of hydroxyl radicals on the surfaces of the delustrant which, with light absorption provides energy for their release and the subsequent initiation of polymer degradation.[6]

Metal ion impurities within textile fibres may also catalyse photodegradation. Transition metal ions in particular are effective in this respect. Their action is considered to arise from the formation of an unstable complex with hydroperoxide functions in the fibre. Subsequent electron transfers create free radicals which, by hydrogen abstraction from the polymer, result in its degradation:-

$$ROOH + M^{n+} \longrightarrow RO\bullet + M^{(n+1)+} + OH^-$$

$$ROOH + M^{(n+1)+} \longrightarrow ROO\bullet + M^{n+} + H^+$$

$$RO\bullet + PH \longrightarrow ROH + P\bullet \; ; \; P\bullet + O_2 \longrightarrow PO_2\bullet \text{ etc.}$$

The presence of metal ions can arise from either the polymerisation catalyst residues, or as adventious pick-up during fibre fabrication. With the natural fibres metal ions arise as a result of "sorption" from either the soil or diet dependent upon whether a plant or an animal fibre. Additional uptake can occur from the prevailing environment or during processing. The catalytic influence of metal ions can be inhibited by the addition of complexing ligands. A fully complexed cation has little or no catalytic activity.

Inherent and inadvertent impurities which contain chromophores, or additives introduced for other purposes (eg. dyes) can all act as possible sensitisers for photo-degradation. In the melt spinning of polyamide, polyester or polyolefin fibres the emergent molten filaments from the spinerette do contact air for a short time while the polymer is at a temperature and in an environment conducive for oxidation. This contact time may be as long as a minute if drawing operations are also included. The result of these oxidations during the spinning is the presence of entities which are photo-sensitive/photoabsorbing.

Examples of possible impurities in chemical fibres include unsaturation in the polymer backbone, peroxide or hydroperoxide residues and carbonyl groups. Many of the above act as sensitisers to photodegradation by absorbing radiation and transferring that energy either to the polymer or to oxygen. Very little is needed to have a significant influence. Stabilisers against photo-degradation can act by one of two basic routes:
(1) inhibit the chemical degradation processes, eg. by either acting as radical scavengers or by destroying the radical "source"
(2) by acting (a) as a UV absorber of the active components of the illumination, with the subsequent release of the energy as here, or (b) as a quencher (Q) of the excited entity, acting in the manner:-

$$D \xrightarrow{h\upsilon} D^* + Q \longrightarrow D + Q^* \longrightarrow Q + h\upsilon'$$

The most significant stabilisers against degradation are:
    hindered amines
    hydroxybenzophenones
    benzotriazoles

Nickel chelates are also UV stabilisers which act by decomposing peroxide residues and then scavenging the generated radicals.

Specific fibre degradation mechanisms

Efficient stabilisation involves the effective retardation of the critical pathways by which property changes arise within fibre structures. With regard to change each fibre type will be unique in terms of their

chemical composition, their morphology (for manmade
fibres) and their histology (for natural fibres). For
the total structure, the mechanical properties will be
strongly affected by the characteristics and changes of
the weakest component. In morphological terms the less
ordered volumes, their particular form and the
stabilising influences upon them of the ordered
components are the weak elements. With the complex
natural fibres the particular histological forms, and the
characteristics of the weakest elements thereof, play
dominant roles in determining the observed behaviour of
the whole.

Cellulose based fibres represent by far the largest
component of the raw materials utilised in textiles,
however celluloses their degradation and stabilisation
have been discussed by others and their
structure/behaviour will not be examined here.
Consideration will be targeted at the natural animal
fibres (wools/hairs and silk) and the manmade fibres
(primarily the chemical fibres).

Wools and hairs. Under the microscope wools or
hairs when examined whole, or following a transverse or
longitudinal sectioning, reveal their particular
histological features which are simply represented in
Fig.6. The core of the fibre is formed from
interdigitating cigar shaped cortical cells. The fibre
core is surrounded by flat plate-like overlapping cells
(cuticle cells) which are ca. 0.3-0.5 µm thick and
approximately square 30-50 µm. This cuticle overlap is
the origin of the locking together in compact masses of
wool or hair fibres following mechanical action when wet.
The outer cuticle cells are covered by an epicuticle, an
exterior resistant membrane. Within the fibre this
membrane is a component part of the cell membrane
complex. The major constituent of the cuticular cells is
the exocuticle which is approximately 100nm thick and
represents more than 60% of the cuticle depending upon
the fibre type. The exocuticle contains a high fraction
of cystine.[8] Within the 5µm diameter cortical cells are
smaller structural filamental units (macrofibrils) of
0.05-0.2µm diam. These filaments are composed of 7.5nm
microfibrils, which are then, themselves, structurally
determined by an arrangement of 2nm protofibrillar
elements, each composed of a 4 chain coiled coil
molecular unit. The presence of these protofibrils is
discernable in transverse sections after staining when
examined under high resolution with a scanning electron
microscope.[9] The cortical cells are bound to themselves
and to the cuticular cells by the cell membrane complex.
A typical composition for a wool or hair would be 86.7%
cortical cells, 10% cuticle cells and 3.3% cell membrane
complex.[10] The cell membrane complex (CMC) which links
the whole structure together is formed from the
plasmalemma membrane of adjacent cells

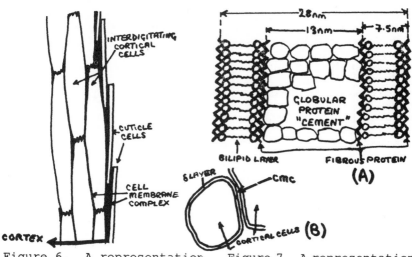

Figure 6. A representation of longitudinal section through a wool fibre

Figure 7. A representation of the structure (A),and location (B) of the cell membrane complex in wools.

together with an intercellular cement. A representation of both its structure and location are illustrated in Fig. 7. Its composition is 0.8% lipid, 1.0% protein and 1.5% resistant membrane. The potential for failure of wools at the cell membrane complex is demonstrated by the filamental surfaces observed in scanning electron microscopic views of fibres failed either in tension or after abrasion (Figs. 12, 13 ref 11). From the structural representation of the CMC in Fig. 7 the potential for a preferred fission plane through the lipid bilayer is apparent. The phenomenon of human hair "splitting" arises from the same structural source.

The current view is that the epicuticle membrane around the cuticle cells provides a barrier controlling the diffusion into the fibre of dyes and other molecules. Leeder[11] suggests that diffusion occurs preferentially between the cuticle cells via the CMC and then into the cuticle and corticle cells. The lipid component of the CMC can be removed by formic acid (24hr at 20°C) or by ethanol. The abrasion resistance in particular is very dependent upon the CMC, and its effective removal increases the abrasion resistance. In water the cell membrane complex swells and even at 100°C the mobile cystine crosslink is converted to a fixed lanthionine covalent crosslink. This conversion has a deleterious effect of the properties expected of wools. Under alkaline conditions the conversion to lanthionine is accelerated. The existence of change/damage can be monitored from the changes in wools solubility in urea-bisulphite. The effects of both temperature and alkalinity on the cystine content/solubility of wools are demonstrated in Fig. 8.

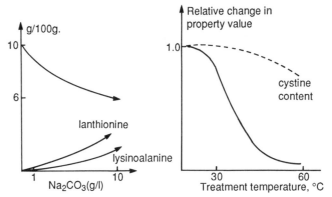

Figure 8. The influence of temperature and alkalinity on the properties of Merino wool.[12]

Under acid conditions the wool protein chains loose the amide residues from their side chains, also acyl N→O migration and peptide bond fissions occur as well as enhanced thiol-disulphide interchange. The occurrence of bond scission can be monitored from the alkali solubility.

Oxidising conditions can result in the conversion of the labile cystine and cysteine residues to cysteic acid.

The potential influence of detergents other than purely lowering the surface tension of the aqueous medium, enhancing the fibre wetting, should not be overlooked. With wools additional physical and chemical influences are possible. Not surprisingly, the bilipid layers are affected as a consequence of detergent presence. These layers are removed or modified to such a degree that detergent treated fibres are more liable to attack (acid hydrolysis) and to damage by enzymes. The wet tensile strengths of wools are reduced if detergent is added to the immersion liquor. Changes in the X-ray diffraction patterns of the ordered elements and their

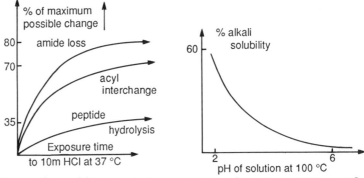

Figure 9. The effect of low pH on the properties of wool.

stability have been described by Spei for wools examined after treatment with sodium dodecyl sulphate.[13] Milligan has also reported the difficulties of total detergent removal from wools after immersion in detergent solutions.[14] The chemical consequences of detergent in very dilute acid conditions can be simply illustrated through the observed increases and decreases in wools alkali solubility. These changes are indicative of there being chemical damage (Fig.10).

With the presence of the outer epicuticle membrane on wools and hairs, the sorption of an anionic detergent by the fibre results in the creation of a lower pH within the fibre than that which prevails in the external bath, and in enhanced acid damage. With cationic detergents an internal higher pH is created within the fibre as a result of the sorption of that agent. No change in solubility is observed with nonionic detergents. If anionic detergents are employed for a simple washing one should note that sulphate half esters such as sodium dodecyl sulphate are slowly hydrolysed by acid, producing sulphuric acid and could cause damage bearing in mind the difficulties of total detergent removal.[16]

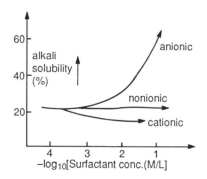

Figure 10. The influence of detergent concentrations in dilute acid on the alkali solubility of wool.[15]

Proteolytic enzymes have the ability to digest stains containing proteins or starches and have therefore been promoted in detergent formulations. Friedman has however observed that enzymes containing detergents damage wools.[17] Fungi and bacteria act enzymatically on the cell membrane complex with consequential fibre destruction.[18] The protein chains of both the inter-cellular cement and of the cortical cells can be cross-linked by formaldehyde action. The mechanical stabil-isation achieved by a mild treatment to introduce but a small number of additional crosslinks is reflected in the fracture profile of treated fibre shown with the scanning electron microscope (ref 11, Figs. 14/15). A potentially convenient vapour treatment on wetted (80% moisture level) wool is described by Khayatt et al.[19]

Treatment of wools with organic solvents can also "damage" subsequent properties. If wools are extracted by any alcohol from methanol up to tert.butanol penetration and extraction of lipid from the CMC occurs. The rates of chemical reaction, including dye uptake, are enhanced following the alcohol treatment, although the limiting extent of action remains constant. The loss of lipid also enhances intercellular adhesion. When chlorinated solvents, as employed in conventional dry cleaning, are used to treat wool textiles, certain problems can arise in the long term. Dry cleaning solvents are lost only slowly on drying wools at room temperature.[20] Morphological changes have been recognised in wools following solvent extraction by a chloroform/methanol soak, followed by a boiling water treatment to remove residues.[21] Microscopy revealed the modification of the CMC and the presence of cuticle damage. After the soak phase alone the wools are more susceptible to hydrolytic damage.[22] Accelerated photo-yellowing of wools pretreated with perchloroethylene has also been reported.[23] If any chlorinated hydrocarbon is retained by the fibres after treatment one should not ignore the fact that the long term photolysis of these solvents can result in some acid products. Treatment of wools by soxhlet extraction with either low or high boiling petroleum ether fractions is claimed to have no effect on wool,[24] although the range of tests applied was not comprehensive.

The base protein chains of wool are constructed from 19 different amino acids. The potential for the photodegradation (including yellowing) must be positively addressed. Photodegradation may result from the absorption of energy quanta at the site of subsequent decomposition/change, or the quantum can be absorbed and transferred through vibrational and segmental motion to a particularly weak bond. Alternatively a separate species may absorb the incident energy and then, either via molecular or segmental collision this energy is transferred directly to the substrate, or the absorbing entity is transformed into an active species capable of reacting with the substrate. With wool the potential exists for the direct absorption with degradation/ destruction at or close to the absorption site. With the exception of tryptophan, cystine residues suffer more photolytic degradation than any other residue in wool,[25] through actions such as: $\vdash CH_2SSCH_2 \dashv \xrightarrow{h_\upsilon} \vdash CH_2SS\cdot + \cdot CH_2 \dashv$. The yellowing of the wool by the transformation of the tryptophan residue to a chromophore has been examined by several workers.[26/27] Tyrosine and histidine residues have also both been implicated in the photo-yellowing of wool.[28] Dyes can also sensitise the photodegradation of wool. The mode of action following the irradiation of the dyed fibre is as follows, (D = dye; WH = wool protein):- D $\xrightarrow{h_\upsilon}$ D* ; D* + WH → DH + W· a reactive wool protein able to undergo further change. DH is likely to be a "colourless dye" which in air is oxidised back to its original form. Tryptophan, tyrosine and histidine are residues sensitive to degradation by this route with there being oxygen absorption and carbon dioxide evolution.

Conservators have recorded observations where silk adjacent to wool in a textile is subject to accelerated photodegradation. Launer[25] analysed the gases evolved following wool irradiation. With high energy "germicidal" ultra violet (254nm) he identified CO_2, CO, H_2, COS, H_2S and CS_2. Such conditions are unlikely to be found in any practical situations, however during irradiation at 365nm only CO and CO_2 were detected while irradiation with 436nm only indicated CO_2. Although the occurrence of wool degradation is apparent, the generation of any gaseous species likely to positively accelerate silk degradation is not clear. The probable major product from the photo-oxidation of cystine or cysteine residues is cysteic acid. The relatively high concentrations of cystine in the cuticle and outer epicuticle, it is suggested, will result, on photolysis, in high concentrations of cysteic acid in the wool outer volumes, sufficient for a lowering of the pH in the adjacent silk to accelerate the degradation of the latter.

To stabilise wools from photodegradation the use of a UV absorber as a protective screen is a useful procedure. Sulphonated triazines such as:-

are suggested. Although photo-bleaching due to absorption in the visible does occur, the retardation is directly related to the screening effect. Because of the possibility of photo-oxidation it is usual to also add an antioxidant (such as a hindered phenol, eg. Irganox 1425 [Ciba Geigy]) to scavenge for the photolytically generated free radicals.

Silks. The other group of protein based fibres is that of the only continuous length (filament) fibre, silk. The average amino acid composition (as mole %) is approximately: 42.9% glycine, 30.0% alanine, 12.2% serine and 4.8% tyrosine, along with 0.9% threonine and 0.67% phenylalanine. The above represent the major identified residues together with tryptophan, lysine and a very small amount of cystine. The major individual protein chains are assessed as of 103,000 average molecular weight; four of these are joined together via 4 cystine covalent bridge links to form a single unit, and a short chain of only 27,000.[29] The structure of the continuous filaments is primarily of filamental fibrous elements. Damaged silks show, at their damage sites, clear fibrillation of the filaments,[30] which suggest a micro-fibril structure. The organisation within such microfibril units are ordered volumes with β protein chains in an extended arrangement. The morphology is shown diagrammatically in Fig. 11. The fibrillar nature of the silk filaments is also the source of a change in the visual appearance to the textile described as "lousiness".

The ordered volumes comprise protein chain segments essentially composed of the less bulky amino acid residues (eg. GLY, ALA and SER). It is recognised that within the fibroin protein chain are alternating GLY-ALA sequences with some of the ALA replaced by SER and occasionally TYR. Suggestions are made that the ordered volume segments are primarily composed of repeating hexameric units GLY.ALA.GLY.ALA.GLY.SER. The bulkier amino acid residues will be the source of, and arise at the disrupted order fringes. The bulkier residues are located therefore in the more accessible, less ordered volumes.

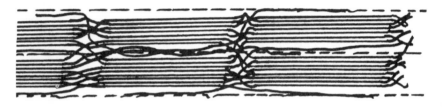

Figure 11. A representation of the molecular organisation forming the fibrillar elements within silk filaments.

In the transformation of the silk as "thrown" from the silk cocoon as a double fibroin filament with a sericin protein gum, to the 100% fibroin textile product, the removal of the sericin results in a significant reduction in weight. As silk is normally traded by weight the Europeans developed a process of "weighting", of restoring the whole of the silk after degumming to approximately its previous mass. This weighting was commonly achieved by the absorption of tin salts, yielding concentrations of SnO_2 within the final product. The added weight also aids the aesthetics of the textile product eg. the drape and hand.

Silk is the most readily photodegraded of all the natural fibres. The action appears to be primarily an oxidation. The need for oxygen was clearly demonstrated by Egerton.[31] Moisture only marginally enhances the rate of photodegradation, but the sensitivity of fibroin towards damage is determined by the pH. The more acid the environment the more rapid is the degradation. In the presence of mineral acid the degradation appears to involve both the serine and threonine residues.[32] Reference the practised tin weighting, it is now recognised that Sn(IV) is an effective catalyst for the photodegradation. On light exposure, tensile strengths fall for silks containing not just Sn(IV) but also Fe^{3+}, Ni^{2+}, Zn^{2+} or Ti(IV), although when Mn^{2+} or Cu^{2+} cations are present the observed strengths rise. Recently devised substitute weighting procedures[33] for the application of metal salts utilise epoxides, and these can concurrently improve the "easy-care" characteristics. Photodegradation is accompanied by a loss in mass, and a fall in the mechanical properties as well as a discernable yellowing. The presence of stains, and acid or vat dyes in silks increase their rate of photo-degradation as do fluorescent whiteners.[34] The latter influence is illustrated in Fig. 12. Not surprisingly, as with wools and hairs any retained dry cleaning solvent will also accelerate photodegradation.

Although tyrosine has long been implicated in the yellowing/degradation of silk, its full role has still not been established. 90% of the TYR in silk is in the accessible volumes and, as photodegradation proceeds with increasing yellowness index, the TYR composition of the

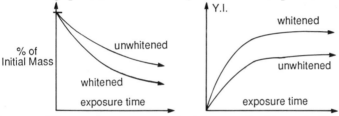

<u>Figure 12.</u>   A representation of the influence of fluorescent whitener presence on the degradation of silk exposed to 365nm illumination.

residual textile shows the most significant fall. The
fraction of order in the textile also appears to increase
with irradiation. The increasing yellowness of silk[35]
follows closely also the fall in tryptophan content.
As silk is exposed to light the serine content remains
unchanged but the tryptophan, although only present in
very small amounts, decreases gradually in content as
yellowing occurs, with the rate being faster in the
presence of tyrosine.[36] In the degradation, fission of
the peptide bond adjacent to TYR has been suggested, as
well as oxidation of the TYR residue itself to a
quininoid structure, ie. a chromophore absorbing in the
visible.[37] No quininoid structure has however yet been
isolated from degraded silks.[38] Ammonia has been
identified as a product of photoaction on silk and
equilibrium regain levels are higher following
degradation. Despite the suggested bond fission at TYR,
other amino acid analysis studies of the products from
photo exposed silks suggest that scissions are random.
At least in the early stages of exposure, there is
observed an increase in the soluble silk fraction.
Extended photo-oxidation of silk can, concurrent with
colour development, render the fibre insoluble in a
solution of calcium chloride and formic acid. This
observation led Earland to suggest that a "melamine like"
crosslink[39] was formed between two oxidised tyrosine
residues. An alternative source of crosslinks could be
the reaction of the oxidised tyrosine residues with $\in$ -
amino group of lysine residues.

    To prevent photodegradation two simple routes
suggest themselves, either the addition of an anti-
oxidant/UV absorber system or the modification of the
amino acid residues at the sites responsible for the
degradation. Where silks contain tin salt the use of
chelating agents to deactivate the influence of the
cation catalyst is effective. Thiourea plus aldehyde
treatments are described as either preventing or
retarding silk photo yellowing. If one considers the
"active site" modification route, Tsukada has treated
silks with glutaric anhydride.[40] The claim is that the
main reaction sites are at TYR, SER and lysine. The
anhydride action introduces a crosslink between two of
the above sites. It is reported that the mechanical
properties are effectively unchanged by the treatment but
following an 8% add on, the yellowing is markedly reduced
after a 100hr Xenon light exposure, compared with the
untreated textile.

    If the silk damage is localised, indicators as to
the degradation paths may be possible. Exposure of the
textile to a sodium zincate solution will result in
localised swelling at the sites of preferential
degradation. When considering methods to recognise the
occurrence of protein based fibre degradation at the
earliest possible stage, one method which could offer
potential assistance is thermogravimetry. Despite the

fact that the sample is degraded in the procedure only 1mg. of fibre is needed. A sample, predried under vacuum, is heated on a balance under continuous evacuation at a slow rate of 1 deg./min. from 150 upto 400°C. The mass and temperature data are transformed into a rate of fractional mass loss as a function of sample temperature which is displayed against sample temperature.[41] In the case of wools and hairs, reactions/changes involving particular amino acid residues prior to examination can frequently be identified, eg. the oxidation of cystine (see Fig. 13A). Shifts from the reference curve reflect sensitively the occurrence of change. With silks, this changed behaviour as a result of prior "degradation" is also clearly seen by comparing observations with those of an undamaged silk (Fig. 13B).

Manmade Fibres. The potential problems with regard to photodegradation of manmade fibres that arise from either the inherent pick up of metal ions, or the formation of structural and/or chromophoric entities during production were mentioned earlier, as was the influence of added delustrant.

When considering specific chemical fibres, the polyamides are particularly sensitive to photodegradation. Their relative stability (represented by nylon 6.6 in comparison with silk), as the observed relative strength losses following equivalent 16 week exposures, is recorded as: silk 85%, semimatt nylon 6.6 55%, bright nylon 6.6 23% and cotton 18%. The intrinsic viscosity of nylon 6.6 sampled from irradiated fabrics falls with increasing exposure time. A marked yellowing is also observed. When the nylon samples were exposed in an acid environment the viscosity fall is minimal. Recalling possible difficulties in the effective interpretation of viscometric data (loc cit.), the carboxyl end group concentrations were also observed to increase with greater exposure levels. Again a reduced change was observed when exposures were made in an acid environment. The influence of fibre morphology is also recognised by positive effects on the observed viscosity and end group

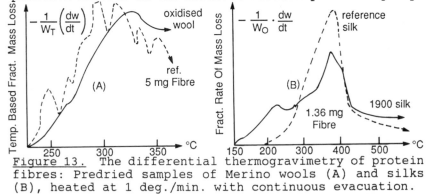

Figure 13. The differential thermogravimetry of protein fibres: Predried samples of Merino wools (A) and silks (B), heated at 1 deg./min. with continuous evacuation.

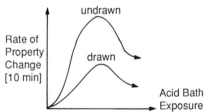

Figure 14. A representation of polyamide fibre property changes observed (intrinsic viscosity, end group analysis) following extended light exposures.

estimations (see Fig. 14). In atmospheres containing oxides of nitrogen polyamides degrade more rapidly. No effects on properties were observed when aliphatic polyamides were exposed to only wavelengths above 290nm in a nitrogen atmosphere. Degradation does occur if the atmosphere is air, and when the illumination is well into the visible. The "degradative" processes continue in the dark, indicative of a radical nature to the significant process. This is confirmed by the stabilising effect of hydroquinone. The favoured mechanism for the photo oxidation involves attack on the CH_2 adjacent to the amide nitrogen:- $\leftarrow CONHCH_2 \rightarrow$ $\leftarrow CONHCHO + \cdot CH_2 \rightarrow$

$$h\upsilon, O_2 \nearrow$$

$$+RH$$

$$\leftarrow CONH_2 + R\cdot \underset{\leftarrow}{\overset{RH}{\quad}} CONH\overset{\bullet}{\underset{O}{CH}} \rightarrow R\cdot \rightarrow CONHCOCH_2^\bullet \rightarrow$$

$$+ \quad HCO \nearrow$$

These and subsequent reactions yield the observed products. Although the inherent stiffer chains of the aromatic polyamides enable their translation to high (mechanical) performance fibres both Nomex and Kevlar are very susceptible to photodegradation. The existence of degradation (strength loss) is noted even following illumination only by radiation well into the visible (500-450nm). Degradation again continues in the dark, supporting a radical mechanism.

The proposed mechanism involves absorption of radiation involving the aromatic entities, with this energy being transmitted to the amide bond which undergoes homolytic fission to radicals. Because of the rigidity of the total molecular structure, only extremely limited segmental mobility can occur and recombination of the primary radicals is the favoured second step. Oxygen can however diffuse into the structure and reactions involving the primary radicals arise such as:-

$$R\cdot + O_2 \longrightarrow ROO\bullet \longrightarrow ROOR$$

$$R^1H \nearrow \qquad R\cdot$$

$$R^1H \qquad R^1H$$

$$R^1_\bullet + ROOH \longrightarrow HO\bullet + RO\bullet \longleftarrow ROH + R^1_\bullet \longleftarrow 2RO\bullet$$

Solvent extraction procedures can remove the transition metal catalyst residues and reduce the sensitivity of the treated fibre to photo-oxidation.

Polyesters are considered by many to be stable, however polyethyleneterephthalate is susceptible to light of less than 315nm, because of absorption by the estercarbonyl. Wiles[42] observed crazing of irradiated polyester fibres with yellowing at the surfaces, which suggest that degradation is localised in the surface volumes. A fall in the observed average molecular weight further suggests chain scission. Wiles et al., also identified a fluorescent product with a spectrum consistent with the identity

$$-\text{O·CO}-\underset{\underset{\text{OH}}{}}{\bigcirc}-\text{CO·O}- \quad \bullet$$

If the primary product from photolysis is the radical R·, then subsequent reaction with molecular oxygen, and hydrogen extraction would produce a hydroperoxide which is a potential source for hydroxyl radicals. The carboxyl concentration as acidic chain ends is observed to rise following irradiation in nitrogen, which implies that the acid function must be a product of the primary photolysis. An increased concentration of acid residues is generated if the fibres are irradiated in air; this increase must be the consequence of an oxidation. For polyethyleneterephthalate fibres contact with, and uptake of dry cleaning solvents results in an enhanced segmental mobility within the less ordered volumes.[43] This mobility facilitates further molecular ordering and some fibre shrinkage, the occurrence of which results in changed properties. Also, as a result of the induced additional order some solvent can be trapped within the structure with the possibilities for its involvement in photolytic processes on subsequent irradiation. Particularly when delustrant is present photolytic degradation of polyesters is noted. Viscosity observations were unsuccessful as a method for assessing degradation extent,[44] but strength losses and end group analyses positively identified that chain scission is occurring.[45]

The polyolefin fibres and polypropylenes in particular positively degrade in oxidising environments. The locus of this oxidative breakdown is the hydrogen of the tertiary carbon. Equivalent sites exist at the branch points of low density polyethylenes. The susceptibility to photo-oxidation increases in the sequence: HDPE < LDPE < PP. The rate of the poly-propylene oxidation is less than one might expect, primarily because of the higher fraction of order observed in these fibres. Linear hydrocarbons are transparent to UV, and so photodegradation must be initiated from either defects which are chromophores (eg. —OOH or $\text{C}=\text{O}$), or metal impurities (particularly catalyst residues). The peroxide defect is the most significant, coupled with catalyst residues able to

promote its homolytic fission. If the polyolefin
contains any unsaturation within its structure,
decomposition arises involving environmental impurities
such as SO_2 and nitric oxides.

Stabilisers for those substrates where a radical
based mechanism dominates are benzophenones or hindered
amines, provided that those chosen are compatible with
the substrate. The catalytic effect of trace cations on
the decomposition of peroxides to radicals and thus also
on the fibre degradation is particularly relevant with
the chemical fibres. If the cation is a strong reducer
(ie. readily oxidised) such as Fe^{2+} and Cu^+, then its
influence as an accelerator of decomposition is
significant. To inhibit this activity the ion requires
to be extracted, or complexed by such as 9-hydroxy
quinoline. This latter procedure is adopted to inhibit
the copper based oxidation of polypropylene. In
polyester fibre significant "foreign" metal ions are
present (eg. Ti at levels 0.01 to 1%, Zn at 10 to 250ppm
and Sb at 20 to 500ppm) and their presence is responsible
for accelerating degradation. With the polycrystalline
chemical fibres, folded chain morphologies are found in
for example the polyolefins and polyesters. The strained
segments of the chains at the fold surfaces and those
within the rest of the low order volumes are particularly
susceptible to degradation. Photo-oxidation processes
must be contained within the less ordered (accessible to
Oxygen) volumes. These same volumes are those in which
added stabilisers will be located. If one considers a
high density polyethylene with 0.75 fraction of order and
with 0.5% W/W added stabiliser, the "effective"
concentration in the susceptible less ordered fraction
will be 2.0%. To reduce the rates of stabiliser loss, be
they anti-oxidant or UV absorber, it is desirable to
increase their molecular size. Such increases, if
appropriately designed, will achieve greater
compatibility with the basic fibre, a slower rate of
diffusion through the substrate and a reduced rate of
loss from the fibre surface.

Conservation.

To consider finally the conservation of the total
textile; several methods are currently practised and
others suggest themselves as worthy of consideration.
The alternatives are identified in table 5.

The possible uses of methods 1 and 2 have been
discussed within the text. Degradation of mechanical
properties is often a consequence of chain scission; if
a limited number of specifically located interchain links
are introduced to effectively "replace" those chain links
lost by scissions, the mechanical features of the whole
can be largely sustained. Excessive crosslinking however
results in an undesirable shift towards a brittle
product. Method 5 relates either to the provision of a

<u>Table 5</u>. Textile conservation methods: Procedures to retard/inhibit degradation <u>and</u> to stabilise the degraded textile.

1. Addition of stabilisers
2. Modification of sensitive sites
3. Controlled crosslinking of chains
4. Consolidation
5. Binding to a support
6. Supporting "screens":
   (a) sprays
   (b) interfacial polymerisation

physical textile support for the now fragile degraded textile elements or the encasement of these elements between layers of a transparent thin polymer film. For this film polyester is frequently suggested. In the application of the latter procedure, even with thin films a relatively stiff product must result implying also a marked deviation from the original textile character-istics. The use of consolidation has its advocates.[46] By sorbing monomer vapour into the textile followed by polymerisation initiated by irradiation offers a facile handling route. However, the irradiation itself may induce further degradation of the textile and, more significantly, the polymerisation involves a radical mechanism. Transfer of the radical activity to the textile substrate is likely, with the subsequent formation of a covalently bound polymer fragment grafted to the substrate.[47] The ideal requirement of 100% reversibility for the conservation method is lost. The levels of consolidation (polymer "add on") are relatively high, and "desirable" textile features are also lost as a consequence.

The use of a prepolymer spray onto the fragile textile with subsequent post-curing offers the potential for providing a physical support to the whole, concurrently with a barrier against oxygen absorption and UV influence. One practical difficulty is the achieving of a uniform level of low add on over the whole textile, and without adverse effect to features like handle. An alternative procedure for achieving a uniform thin supporting and barrier providing film is worth further examination. Interfacial polymerisation has long been practised as a means of providing a fine polymeric film totally covering all fibres to achieve machine washable wool textiles. A low thickness film (a couple of microns) is achieved together with some "spot welds" between yarns, without detriment to the textile handle. The basic principle is to employ a condensation polymerisation in which one reaction component is presented in an aqueous phase with the second reactant in an immiscible organic phase. The specific reactants, if appropriately chosen, can avoid potential involvement of functional groups from the fibre. With the aqueous phase can be included an antioxidant and/or UV stabiliser. The

textile is first allowed by immersion to take up the
aqueous component (plus stabiliser) and is then, after
removal of excess liquor, immersed in the organic phase.
Polymerisation will arise at the interface between the
two phases, which is the fibre surface. Appropriately
controlled, a fine film is formed on the surface of the
fibre components of the textile with the stabilisers
trapped within. This film with spot welds acts as the
mechanical support for the whole, as well as providing a
barrier to oxygen ingress and active light. Trials of
this technique have been conducted with both silk and
cellulose substrates, however more needs to be done to
identify any potential problems, and to optimise the
treatment procedure.

REFERENCES:

1.   K. Valinejad, MPhil Thesis, CNAA, 1990.
2.   A. Keller, Macromol. Chemie, 1971, 141, 189.
3.   R.J. Elema, J.Polymer Sci (Symposia) 1973, 42,
     1545.
4.   L.M. Lock and G.C. Frank, Text. Res. J., 1973, 43,
     502.
6.   Chapter 5 in N.S. Allen and J.F. McKellar (eds)
     'Photochemistry of dyed and pigmented polymers'
     Applied Sci. Publ., London, 1980.
7.   J.A. Swift in 'Chemistry of natural protein fibres',
     R.S. Asquith (ed), Wiley and Sons, London, 1977.
8.   J.H. Bradbury, Adv. in Protein Chemistry, 1973, 27,
     111.
9.   J.A. Maclaren and B. Milligan, 'Wool Science',
     Science Press, NSW, 1980.
10. H. Zahn, 6th Int. Wool Text. Res. Conf., 1980, 1,
     supplement.
11. J.D. Leeder, Wool Science Review No. 63, 1986.
12. K. Ziegler, in ref. (7).
13. M. Spei, Appl. Poly. Symp., 1971, 18, 659.
14. B. Milligan, Text. Res. J., 1989, 59, 653.
15. J.C. Dickinson, Wool Science Review No.51, 1975.
16. J. Steinhardt and C.H. Fugitt, J. Res. N.B.S., 1942,
     29, 315.
17. M. Friedman, Text. Res. J., 1971, 41, 315.
18. F. Kidd in ref. (7).
19. S.M. Khayaat, H.L. Needles, S.A. Suddiqui and S.N.
     Zeronian, ACS Symp. 260 'Polymer for fibres and
     elastomers', J.C. Arthur (ed).
20. L.A. Holt, Text. Res. J., 1984, 54, 226.
21. L. Jones, Text. Res. J., 1989, 59, 530.
22. D.E. Rivett, J. Text. Inst., 1989, 80, 261.
23. M.P. Mansour, H.J. Cornell and L.A. Holt, Text.
     Res. J., 1988, 58, 246.
24. M. Soturiou-Provata and A. Vassiliadis, Text. Res.
     J., 1966, 36, 1031.
25. H.F. Launer and D. Black, Appl. Poly. Symp., 1971,
     18, 347.

26. A. Meybeck and J. Maybeck, <u>3rd Int. Wool Text. Res. Conf.</u>, 1965, <u>2</u>, 525.
27. L.A. Holt and B. Milligan, <u>J. Text. Inst.</u>, 1976, <u>67</u>, 269.
28. J.A. Maclaren, <u>Text. Res. J.</u>, 1963, <u>33</u>, 773.
29. C. Earland and S.P. Robbins, <u>Int. J. Protein Res.</u>, 1973, <u>5</u>, 327.
30. K. Mahall, <u>Textiles Asia</u>, 1985, no.10, 95.
31. G.S. Egerton, <u>J. Text. Inst.</u>, 1948, <u>39</u>, 293.
32. A.S. Tweedie, <u>Canad. J. Res.</u>, 1938, <u>16</u>, 134.
33. Y. Tanake and H. Shiozaki, <u>7th Int. Wool Text. Res. Conf.</u>, 1985, <u>4</u>, 441.
34. Proc. Amer. Assoc[n] of Text. Chem. and Col., 1989, 37.
35. I.H. Leaver and G.C. Ramsey, <u>Photochem Photobiol.</u>, 1969, <u>9</u>, 531.
36. Z. Yoshida and M. Kato, <u>J. Chem. Soc. Japan</u>, 1955, <u>58</u>, 274, 667.
37. D.B. Das and J.B. Speakman, <u>J. Soc. Dyer and Col.</u>, 1950, <u>66</u>, 583.
38. D.A. Sitch and S.G. Smith, <u>J. Text. Inst.</u>, 1957, <u>48</u>, 341.
39. C. Earland and D.J. Raven, <u>6th Int. Wool Text. Res. Conf.</u>, 1980, <u>2</u>, 173.
40. M. Tsukada and H. Shiozaki, <u>J. Appl. Poly. Sci.</u>, 1989, <u>37</u>, 2637.
41. J.S. Crighton and P.N. Hole, in "Thermal Analysis" (H. Chirara ed), Kagaku Gijutsu-sha, Kyoto, 1977, 301.
42. M. Day and D.M. Wiles, <u>J. Applied Poly Sci.</u>, 1972, <u>16</u>, 191.
43. E.L. Lawton and D.M. Cates, <u>Text. Res. J.</u>, 1978, <u>48</u>, 478.
44. A.H. Little and H.L. Parsons, <u>J. Text. Inst.</u>, 1967, <u>58</u>, 449.
45. S.H. Zeronian, K.W, Alger and S.T. Omaye, <u>Proc. Int. Clean Air Congr.</u>, 1971, 468.
46. see N.H. Tennent, <u>Rev. Prog. Coloration</u>, 1986, <u>16</u>, 39.
47. S. Lenka, <u>J. Macromol. Sci. Rev. Macromol. Chem. Phys.</u>, 1983, <u>C22(2)</u>, 303.

Degradation and Conservation of Cellulosics and their Esters

Jeanette M. Cardamone*, Katherine M. Keister, and Amir H. Osareh

DEPARTMENT OF CLOTHING AND TEXTILES, VIRGINIA POLYTECHNIC INSTITUTE AND STATE UNIVERSITY, BLACKSBURG, VIRGINIA 24060, USA

1. HISTORIC USES OF COTTON AND FLAX

The cellulosic fibers: cotton, flax, rayon, and acetate have interesting and various histories. Cotton may have been grown as early as 2000 B.C. in Middle Egypt before flax was known. India is generally recognized as the center of cotton production since as early as 1500 B.C. until the early 16th century[1,2]. Throughout antiquity to the present, cotton's customary use has been for clothing and for textiles associated with the household. Archaeological cotton fragments used to clothe the dead have been excavated from burial sites. Those that have survived had been exposed to dry and fairly anaerobic environments. The conditions of these textiles in ethnographic collections vary from strong and resilient to weak and friable.

When the chemical changes in cellulose textiles from natural and artificial aging were compared by nondestructive infrared spectroscopy, the spectra showed similar absorption bands. The correlation of these chemical changes with tensile strength loss among the natural and artificially aged samples suggested that artificial aging simulates natural aging.[3]

Flax was distinctively used for religious purposes in Egypt. There are accounts that the Hebrews believed linen garments were worn by angels and that they used fine twilled linen to veil their tabernacle doors.[4] The Roman Empire was said to be famous for its linens of great fineness and whiteness.[5] The oldest prototypes for natural aging are mummy linens with some being as old as 2700 years. These naturally aged linens when extracted from the middle layers of mummy wrappings generally appear in good condition but yellowed with age.

When these linens were analyzed with infrared

---

*Current Address: Eastern Regional Research Center, U.S. Department of Agriculture, Philadelphia, PA

spectroscopy the extent of degradation could be estimated by the selective absorptions of certain functional groups in cellulose.[6]

Natural cellulosics maintained their preeminence in clothing through the 18th century indienne craze as lawn, batiste, muslin, gauze, chintz, and calico fabrics. With the invention of the power-driven knitting machine in 1863, the famous cotton union suit made it possible for men to switch from the drop-seated woollen underwear. One needs only to peruse the 1900 Sears, Roebuck and Co. Consumers Guide to assess the popularity of cotton apparel in bathing suits, shirts,dresses, underwear and drawers. Linen maintained the aura of "cleanliness" for the Renaissance man who often exposed the clear, fresh, detailed border on a linen collar to emphasize the neckline of a person above the industrial ranks.[7] Throughout this period linen continued to be the mainstay for household use, for underclothes and all forms of linings. From medieval times, linen has been the preferred table cloth. The damask outer cloth was prized as most important and was so displayed to depict the status of the owner.[8] Costume collections are also replete with delicate dresses of handkerchief linen from the 1900's.

2 DEVELOPMENT OF RAYON AND ACETATE

Cotton and flax are natural cellulosic fibers. Rayon is regenerated cellulose. Comparatively, rayon has a fairly recent history. It became prominent as a textile fiber in the last century and the first decade of this. Rayon fibers are made by the viscose process and the term "rayon" was later used frequently to describe cellulose acetate as "acetate rayon". By definition, rayon is "a manufactured fiber composed of regenerated cellulose in which the substituents have replaced not more than 15% of the hydrogens of the hydroxyl groups"(Figure 1).[9] One inspiration for this development was the attempt to simulate silk. Thus, the starting material was wood pulp linters, the cellulose found in the leaves of the mulberry and oak leaves fed to silkworms. The wood pulp for rayon was treated with sodium hydroxide, xanthated with carbon disulfide, and dissolved in sodium hydroxide to form a spinning solution from which cellulose was regenerated by the action of an acid. The spinning process involved extrusion of viscous filaments into the bath and through a spinnerette which resembled the regimen of the spinning silkworm.

The development of rayon from the turn of the century was a difficult struggle. Rayon was inferior to silk in strength and resiliency and when wet, it lost even more strength and lacked dimensional stability. The regenerated cellulose did not exhibit the same desirable properties of cotton and flax and this was due to the entirely different orientation of the molecular chains after regeneration.

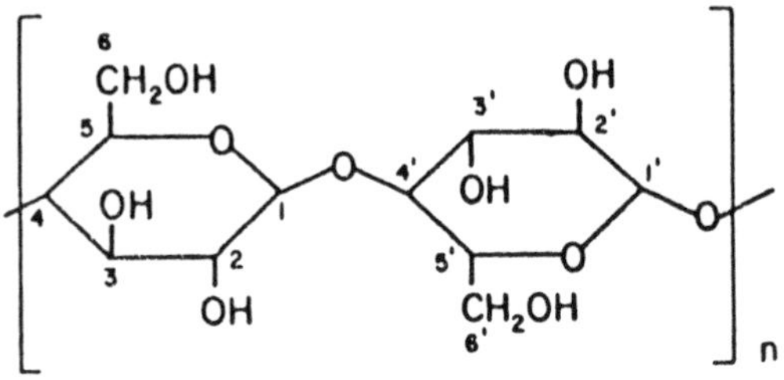

Figure 1   Poly [1,4-beta-D anhydroglucopyranose]

The early rayons were improved by modifying and standardizing the processing procedures so that eventually an acceptable rayon was produced. By the 1940's a variety of rayons with various strengths were developed to meet apparel and industrial end uses. The earliest rayons were filament form to simulate silk but gradually spun rayons were also manufactured to simulate cotton. The commercial availability of rayon in the 1920's was timely for the new flapper fashion rage. The richly dyed rayon fabrics found their way into slinky apparel in shimmering satin, cloche hats and fringed scarfs, beaded apparel, slip-like dresses, rich brocades, beaded chiffons, and the velvets of the 1930's. Lounging pyjamas of blended 50% silk and 50% rayon, machine lace, and evening coats of rayon were popular.

Acetate is defined as "a manufactured fiber in which the fiber-forming substance is cellulose acetate".[10] Cellulose acetates made up to and including the early years of this century were triesters and as such, insoluble in acetone. Partial deacetylation by acid hydrolysis led to soluble acetate (diacetate) which was used during World War I for waterproofing the fabrics covering wings and fuselages of airplanes. After the war, attention was drawn to using acetate as continuous filament yarn by Henry and Camille Dreyfus, the founders of the Celanese Companies. Production of acetate as a second manmade fiber began in the United States in 1924.[11]

Acetate is modified or secondary cellulose; the ester form where two of the three hydroxyl groups on cellulose are esterified. In its production, purified cellulose from wood pulp or cotton linters is mixed with glacial acetic acid, acetic anhydride, and a catalyst. The mixture is aged for partial hydrolysis and is precipitated as acid-resin flakes. These flakes dissolve in acetone and the filtered solution is spun by extrusion through a spinnerette enclosed within a chamber of warm air for acetone evaporation. The formed filaments are stretched and wound onto bobbins. Esterified cellulose was the first thermoplastic fiber. Consumers experienced its heat-sensitivity and decreased moisture absorption. Acetate was used in apparel for shirts, sportswear, and linings. As a simulator of silk, like rayon, it was suitable for the latest fashion designs.

3 CELLULOSE CHEMISTRY

Many factors influence the conditions of textile artifacts. The previous history of use and the ability to resist wear are actual determinants. Many component factors such as fiber and yarn type, coloration and finish applications, fabric weight and construction influence how a textile will survive use in wear and care. The underlying determining factors for estimating condition lie ultimately with the chemistry of the fiber.

Cellulose is a naturally occurring linear polymer, a polysaccharide, formed from glucose, the source of anhydroglucose units linked by the 1,4-beta glycosidic bonds. The basic polymeric repeat unit is shown in Figure 1. This polymer has been characterized as having a viscosity-average molecular weight of 800,000 and a degree of polymerization of 5,000.[12] The salient structural features are the sites of glycosidic bonding between anomeric C1 and O where hydrolysis can occur in an acid environment. Cleavage at these sites form reducing ends from pyranose ring openings and subsequent aldehyde formation. Polymer chains grow shorter and strength decreases. The reaction may proceed by peeling or unzipping from the reducing end, in, to shorten the chain. The rate and extent of hydrolysis depend on the type of acid in the environment, the temperature, and the accessibility of the cellulose to the reactants.

Cellulose can degrade by oxidation primarily at the sites of hydroxyl groups on C2, C3, and C6. In Figure 2, the points of attack from oxidation include C6 to form aldehyde (II), further oxidation of the aldehyde to carboxylate in (III), and oxidation of C2 and C3 to ketone in (IV). Aldehyde groups on C2 in (V) will oxidize to carboxylate in (VI). Even though chain units can be oxidized, the gylcosidic bonds joining the anhydroglucose units at C1 and C4 in the macropolymeric chains can remain intact. Oxycellulose formation causes disruption of the hydrogen-bonded inter- and intramolecular network which binds cellulose in a rigid lateral array of parallel polymeric sheets. Strength will thus be reduced and more accessible regions within the fiber will become available for attack. Oxycellulose formation can occur naturally with time under ambient conditions. This process can be accelerated by artificial heat aging where oxycellulose products form quicker to permit kinetic

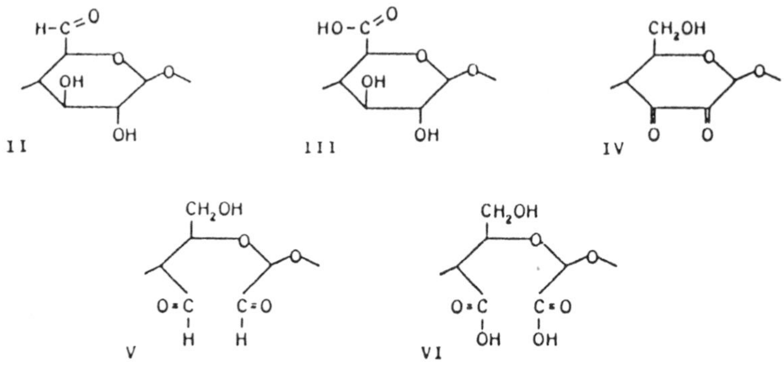

Figure 2   Oxycellulose forms of 1,4-beta-D anhydrogluco-
pyranose

monitoring by sensitive instrumental methods such as
infrared spectroscopy for detection of characteristic
carbonyl and carboxylate absorption frequencies from 1600
to 1750 cm^{-1}.[13]

4. INFRARED SPECTROSCOPY

Infrared spectra were recorded on a Mattson Sirius 100
Fourier Transform Infrared (FTIR) interferometer at 2 cm^{-1}
resolution. The spectrometer was fitted with a liquid
nitrogen cooled mercury-cadmium-telluride detector with
transmission window 4000 to 500 wavenumbers. Interfero-
grams were collected over 8,000 scans. A Mattson horizontal specular reflection sample cell holder of fixed
angle, 30 degrees incidence was used. Samples were
analyzed nondestructively by laying the fabric across the
opening at the top of the sample cell holder. The spectra
in Figure 3 A,B,C,D,E,F,G are the difference spectra of
the unheated cotton cloth minus cotton print cloth (3.8
oz/yd.sq., 80x80 fabric count, from Testfabrics in
Middlesex, New Jersey, USA) heat aged at 190°C for
2,4,7,11,14,23, and 31 hours respectively, in air. The
spectra in Figure 4 A,B,C,D,E,F,G are the difference
spectra of the unheated cotton cloth minus cotton cloth
heat aged for 2,4,7,11,14,23, and 31 hours respectively,
in nitrogen. The carbonyl and carboxylate absorptions in
the 1600 to 1750 reciprocal centimeter absorption regions
were measured for all these samples. Since the same size
and type of sample were presented to the infrared beam
for each data set collection, there was no variation in
sample size and the absorptions could be measured as the
depth of the valley formed in this absorption region. In
Figures 3G and 4G, heat aging after 31 hours in both air
and nitrogen was most pronounced. This absorption region
for the nitrogen samples in Figure 4 A-F appears to be
appreciably less than for the air-aged counterparts.

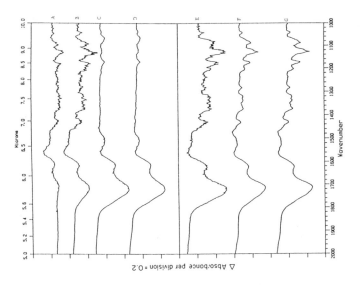

Figure 3 A-G   Difference Spectra: Aging of cotton print
cloth in air at 190°C for 2,4,7,11,14,23,31,hours, A-G

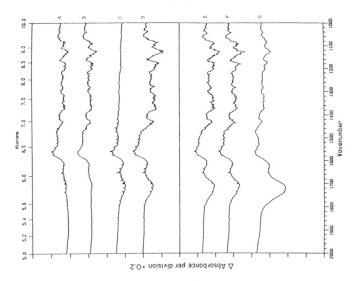

Figure 4 A-G   Difference Spectra: Aging of cotton print
cloth in nitrogen at 190°C for 2,4,7,11,14,23,31 hours,
A-G

The graphic representation of these results can be found
in Figure 5 where after 7 hours of aging in air, cellulose appears to have reached a "levelling off point".
After seven hours, no more of the hydroxyl groups may be
available for reaction. This concept has been related to
the extent of amorphous regions within the polymer.[14]

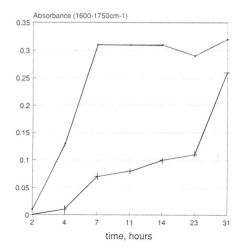

Figure 5 Carbonyl, Carboxylate absorption region, 1600-1750 cm^{-1} versus time for heat aged cotton cloth at 190°C in air . and in nitrogen +

In Figure 5, oxycelluloses form at a constant rate in the initial stages of aging. This profile would ensue in natural aging. The extent of aging in 2700 year old mummy linen was found to be approximately the same as 4 hours of aging in air at 190°C.[15] The natural aging process is quite slow under usual ambient conditions. The implication is that there is time to intervene with treatment to slow, arrest, or reverse the phenomenon provided the long-term future effects of the treatment are known and accepted. Since little is known in this area, perhaps the best prescription is to conserve a cellulosic textile artifact under nitrogen. After 31 hours of aging in nitrogen, the extent of degradation is still less than after 7 hours in air and the levelling off point has not been reached.

It has been shown that thermolysis of cellulose below 300°C leads to weight loss and depolymerization. Thermal analysis and kinetic studies indicate the mechanism is by oxidation and the products include hydroperoxide, carbonyl, and carboxyl functional groups, carbon monoxide and carbon dioxide. The experimental results presented here by IR spectroscopy are consistent with the autooxidation mechanism involving free radicals which progress through three stages: initiation, propagation, and decomposition. It has been shown that below 300°C, the reaction in air is faster than in the absence of air based on carbon dioxide and carbon monoxide production. It has been assumed that these two gases are formed by decarboxylation and decarbonylation, respectively as well as by other competing reactions.[16] At low temperature accelerated aging (less than 190°C) it appears that the

initiation period for cellulose free radical degradation
is prolonged.

5 REACTIVITY AND MORPHOLOGY

It is important to note that the degradation of cellulose
fabric is heterogeneous, because, the substrate is a
solid. The reactions are topochemical and are limited to
the accessible amorphous regions as noted above. Cellulose in its many forms can have a wide range of accessib-
ilities. When measured by the hydroxyl-group hydrogen
deuterium exchange method, the order of accessibility to
moisture was rayon > finely ground cotton > mercerized
cotton > bleached cotton.[17] The hydrolysis-oxidation
method for quantitative determination of accessibility
showed viscose rayons, 25% to 35% accessible and high-
crystallinity rayons, 10% to 15% accessible. Cotton
celluloses were 5% to 10% accessible.[18] This indicates
little of the available cellulose is reactive. Accessibility is inversely related to crystallinity which has
been reported from x-ray diffraction studies as 70% for
cotton and flax, and 40% for viscose rayon.[19] The crystalline content of rayon is lower than cotton and the degree
of polymerization is low at approximately 500 with much
shorter polymer chains. Although the degree of order in
viscose rayon can be improved by the extent to which the
polymer chains are stretched or drawn during spinning,
low crystalline rayon has lost the fibril morphology of
cotton and the low degree of order is responsible for low
strength.

When cotton and flax are compared, cotton has a well-
defined fibril structure with primary wall comprised of
multifibril layers spiralling 70 degrees across the fiber
axis in the primary cell wall and spiralling 20 degrees
to 30 degrees across the fiber axis in the opposite
direction in the secondary wall. Although cotton is
highly crystalline, it is unoriented in relation to the
fiber axis. By contrast, flax fibers are highly crystalline and oriented with polymer chains aligned along the
fiber axis. This confers high strength and rigidity along
with the encrusting matrix found intimately associated
with the ultimate flax fiber cells. These bundles of
cells are consolidated like a composite in the middle
lamella substances known as lignin and hemicellulose.
They represent up to 10% by weight after retting and
possibly 2% after bleaching. Bleaching to the point of
freeing the ultimate cells is not recommended as the
fibers will then be short and weak. Thus a certain amount
of lignin is present in flax even in the bleached state.
The lignin is naturally a brown color and it is bleached
along with the fiber. It is recommended that only the
surface be delignified so that the inner parts of the
fiber are not affected appreciably. The structural model
of lignin is found in Figure 6.

The mechanical properties of flax reflect the interaction between the ultimates and the middle lamella substances. Fibers with significant lignin content have a tendency to yellow or brown on the surface with prolonged exposure to light. The color is due to oxidation of lignin from photosensitivity.[20] The pathway or yellowing by lignin is through the phenolic groups. In the process of yellowing, chromophore groups such as quinonemethide moieties form. These quinoids can oxidize cellulose and hemicellulose components and thus become reduced to hydroquinone derivatives. Since the hydroquinones are leuco chromophores, in the presence of oxygen or peroxide, they are oxidized back to quinone chromophores. Etherification or esterification may be a pathway to arrest yellowing such as occurs in photosensitization.

Acetate fibers contain less crystallinity than rayon by virtue of the inability of the acetyl groups to hydrogen bond as in unmodified celluloses. The attractive forces between polymeric chains are thus weaker. Bulky acetyl groups prevent close packing into a crystalline lattice. Acetate fibers are weaker and have a softening point near 190°C to 205°C and a melting point at 260°C.[21] They are more extensible than viscose rayon fibers.

6 DEACIDIFICATION STUDIES

Various studies have addressed the problem of acid formation with aging in cellulosics. Some have examined artists' watercolor pigments for color changes upon deacidification with barium hydroxide, calcium hydroxide, calcium bicarbonate, magnesium bicarbonate, and methyl magnesium carbonate.[22] Deacidification treatment was found to be detrimental to the natural colorants: logwood, litmus, tumeric, and cochineal and the effects were pH dependent.

In research to stabilize paper strength against the affects of natural aging, the surface was chemically modified to form acetyl cellulose in one case and in another case, cellulose was graft copolymerized with methyl methacrylate. Graft copolymerization was one way to block the hydroxyl groups to prevent the formation of chromophoric ethylenic carbonyls which form from aging by dehydration and the formation of carbonyl groups. Both treatments stabilized paper somewhat against discoloration.[23]

Another study involved the heat aging of deacidified cotton cloth at 150°C and low relative humidity to study

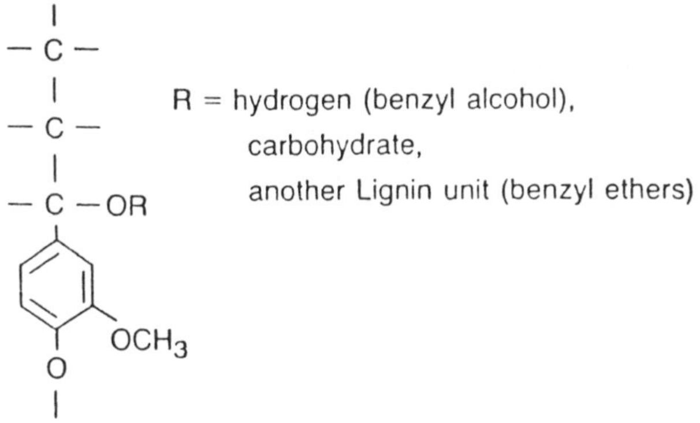

$$
\begin{array}{c}
| \\
- C - \\
| \\
- C - \\
| \\
- C - OR \\
\end{array}
$$

R = hydrogen (benzyl alcohol),
carbohydrate,
another Lignin unit (benzyl ethers)

OCH₃

O

Figure 6   Structural model of lignin

the effects of neutralizing oxycelluloses as they formed. Treatments with calcium hydroxide were as effective as simple washing for slowing degradation.[24]

Others showed that a finish of magnesium bicarbonate applied to cloth aged at 100°C at 100% relative humidity was as effective in retarding the rate of degradation in cotton fabric as aging the samples in a nitrogen environment.[25]

In a previous study by Cardamone et al., degradation in museum textiles was evaluated using property kinetics and the Arrhenius extrapolation to room temperature. "Time left" as a viable textile was estimated by accelerated aging in air in the presence of 50-lux tungsten illumination and 55% relative humidity. In the presence of light the first-order rate constant based on yarn strength increased by 1.9 and decreased the half-life by 55.5% Tensile strength was proposed as a parameter to use comparatively to evaluate the future effects of treatments designed to slow, arrest, or reverse aging.[26]

In studies of artificial aging of linen canvas in air and in nitrogen, in light and dark, photosensitization was explained by the light excitation of lignin and pigment materials and their subsequent conversion to free-radical forms which, in the presence of oxygen, cause the formation of reactive cellulose peroxy radicals. Antioxidants and buffers were coated onto linen before exposure to daylight and the results were evaluated by tensile strength. The buffers, ,magnesium bicarbonate and potassium hydrogen phthalate used to protect against acid hydrolysis from hydrogen ion generation by carboxylate formation showed little effectiveness. This raised the question of the role of hydrolysis in photosensitization.[27]

Overall, some studies indicate that cellulose fabrics when treated with alkaline buffers followed by heat aging slow the rate of degradation as monitored by color change, strength loss, extent of oxidation, and degree of polymerization. These treatments must, however, be influenced by the dyes and finishes present and their fastness or reactivity. When these reagents are applied from a water medium, the high moisture regain of cellulosics promotes the swelling necessary for greater accessibility of the reagent to the hydroxyl and carbonyl groups present. It may then be possible to neutralize any hydrogen ions which cause hydrolysis and strength loss. Overall hydrolysis and oxidation of cellulose should be retarded. These effects were comparable to storage in a nitrogen environment. But further tests are needed, especially concerning the change in the textile's physical properties and the reversibility and renewability of treatments.

7 PHYSICAL AND MECHANICAL PROPERTIES

Over time the materials of fashion were fabricated to meet the ever-changing, "in vogue" aesthetic. The desired fabric handle, drape, performance, neatness retention, strength, flexibility, resiliency, and comfort are a function of the fiber's mechanical and physical properties. Fibers and yarns determine in part the geometric qualities of fabric construction. These qualities and a fabric's failure are most fundamentally described in terms of the fiber properties.

The properties of cellulosic fibers are described in Table 1. The moisture regain measured under standard conditions at 65% relative humidity and 20°C indicate that all cellulosics listed absorb a significant amount of moisture. This will influence comfort favorably but will promote microbiological decay. Acetate with the lowest regain will be the least susceptible. It is generally recommended that the air for storage, exhibition, and display be maintained at 55% relative humidity and 20°C.[28]

In Figure 7, the influence of relative humidity on regain measured under standard conditions, indicates that all cellulosics absorb a significant amount of moisture as it becomes available. At 80% relative humidity the rate of water absorption rises steeply with viscose to the highest level at nearly 40%.

In Table 1, the wet strength of rayon is 50% to 60% of the dry strength. Subjecting rayon to environments where moisture regain is high requires careful handling and the necessary means to support the fabric so it will not hang under its heavy, wet weight. The order of resiliency for cellulosics is acetate > cotton > rayon > flax. The cellulosics as a group have poor resiliency.

Table 1   Properties of Cellulosic Fibers

Fiber	Tenacity at Break (g/den) Dry	Wet	Extension at at Break (%) Dry	Wet	Elastic Modulus (g/den) Dry	Wet	Moisture Regain at 65% RH, %
Cotton	3-5	3-6	4-9	5-12	40-90	30-60	7.0-8.0
Flax	2.6-7.7		2.7-3.3				12.0
Rayon	1.5-4.0	0.8-2.5	10-30	15-40	40-70	15-35	11-13
Acetate	1.0-1.5	0.7-1.1	25-40	30-45	25-40	20-35	6.4

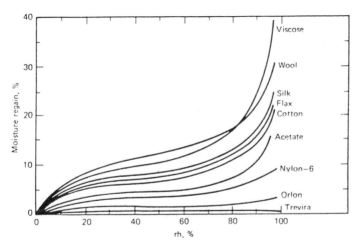

Figure 7   The change in moisture regain with relative humidity at 20°C for textile fibers.[29]

Bending and twisting in care and handling will cause the fabric to fail to return to its original shape.[30,31]

Also in Table 1, the properties of the dry and wet cellulosic fibers differ. Water acts as a plasticizer and helps provide the mobility for chains to slip past each other and align parallel to the fiber axis for greater support of applied stress or load. Cotton when wet is stronger and tougher. It is better able to withstand shock and impact. Wetting rayon softens the fiber, weakens it, and makes it more highly deformable. These effects in acetate, the weakest cellulosic, are more pronounced.

In Figure 8, the mechanical behaviors of dry and wet acetate are compared. Wet acetate becomes softer with a slightly lower elastic modulus and requires less work to rupture. Wet cleaning should be avoided.

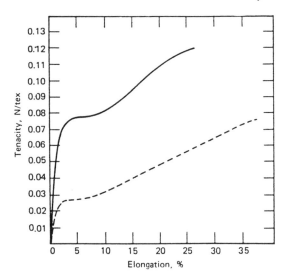

Figure 8   Stress-Strain curves for acetate yarn[32]
at 65 % relative humidity and 20°C: ____
and at 20°C when wet: -----

The mechanical properties of tenacity and elongation are shown by the stress-strain curves in Figure 9

Figure 9 depicts the information in Table 1. It
indicates that flax and cotton are relatively hard and
strong compared to rayon and acetate which are soft and
weak. The applied stress of handling will induce strain
or extension. Moderate extension is 2%. By Figure 9, at
2% extension for rayon and acetate, the fibers' yield
points will not be reached. Flax and cotton have yield
points at zero stress and zero strain. Rayon and acetate
have yield strains at 3% and 4% respectively. Beyond
the yield point there would not be complete recovery. As
seen in Figure 9, both rayon and acetate become permanently deformed at low levels of applied stress. In the
case of cotton, small strains leave significant permanent
deformation.

To further illustrate the mechanical differences
among these cellulosic fibers, Figures 10 and 11 show the
elastic recovery in terms of stress and extension,
respectively.

In Figure 10, permanent deformation from stress is
shown for all cellulosics at stress levels below 10 g
wt/tex. The 30% recovery of cotton under maximum stress
before the breaking point is moderate to poor. Acetate
has the poorest recovery. Both rayon and acetate have
high recovery at stress less than 2 g wt/tex. With the
application of increasing stress the recovery falls off
rapidly at less than 10 g wt/tex. Flax, the hardest and
stiffest fiber,

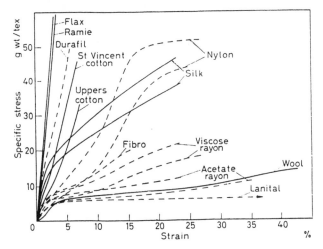

Figure 9   Stress-Strain curves for fibers at 65% relative humidity.[33]

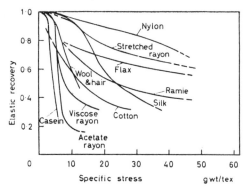

Figure 10   Elastic Recovery versus Stress for textile fibers.[34]

maintains a fairly constant recovery over a wide stress range and can support relatively high stress up to approximately 40 g wt/tex before the breaking point. Fabrics are subjected to stress or applied load when they are lifted or hung and when they sag under their own weight without supports. Permanent deformation can produce fabric rippling and in some cases can cause serious fabric distortions which detract from aesthetic appreciation. Rippling can cause shadowing and dark spots which give the effect of discoloration and in some historic tapestries, the effect is that of premature aging in the faces of mythological figures.

In Figure 11, flax shows the least extension and the shortest recovery range. At best it can recover 80% at 2% extension. Of all the cellulosics, flax is the most dimensionally stable and acetate is the least. A 5% extension results in approximately 44% recovery for

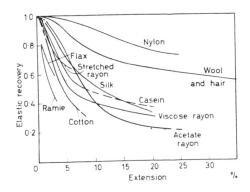

Figure 11   Elastic Recovery versus Strain for textile
fibers [34]

cotton, 55% for rayon, and 80% for acetate. Beyond 10%
extension the elastic recovery of acetate is poorer than
rayon. From this point up to the respective breaking
points, the elastic recovery of both fibers is below 40%.
Great care must be taken to support rayon and acetate
fabrics when they are moved, especially those which are
heavily beaded and sequined or otherwise heavily embel-
lished and are already in a precariously "loaded" position.

8 CONCLUSION

Cellulosic textiles degrade by a complex series of
reactions which begin at the fundamental level of fiber
chemistry. The degradation effects become manifest by the
failure of some measurable macroscopic property. Those
who study aging follow these effects by monitoring such
a property while assuming it subsumes all of the effects
of aging. Tensile strength has been shown to be such a
property. The drawback for its use is that nondestructive
methods are required for estimating a textile's condition. Reflection absorption FTIR is such a method. It can
be brought on site for in situ examination. Estimation of
the "condition age" of a textile can be made by indirect
correlation of the results of this method with tensile
strength measurements made on sympathetic modern heat
aged fabric. More work should be done in this area to
tailor specific conservation treatment to the condition
of a textile's weakest areas. Such methodology would also

provide the means to evaluate the long-term future effects of treatments such as deacidification designed to impede degradation.

Acknowledgements

The authors express appreciation to John Wiley and Sons for permission to use Figures 7 and 8 and to The Textile Institute for permission to use Figures 9, 10, and 11.

REFERENCES

1.   A.N. Galati and A.J. Turner, J. Textile Inst., 1929, 20, T1.
2.   R.F. Nickerson, "Matthew's Textile Fibers", H.R. Mauersberger, ed., Chapter III, Wiley & Sons, N.Y., 194, 102.
3.   J.M. Cardamone, J.M. Gould, and S.H. Gordon, Text.. Res. J., 1987, 57, 236.
4.   W.F. Leggett. "The Story of Linen", Chemical Publishing Co., N.Y., 1945, 38.
5.   Ibid., 45.
6.   J.M. Cardamone, In Proceedings of the "2nd International Conference on Non-Destructive Testing, Microanalytical Methods and Environment Evaluation for Study and Conservation of Works of Art" 1988, Perugia, Italy, II/2.1.
7.   M. and A. Batterberry, "Fashion the Mirror of History", Crown Publishers, N.Y., 1977, 94.
8.   P. Clabburn, "The National Trust Book of Furnishing Textiles", Penguin Books Ltd., London, 1988, 129.
9.   Federal Trade Commission "Rules and Regulations Under the Textile Fiber Products Identification Act", 1986, Federal trade Commission, Washington, D.C., 6.
10. Idem.
11. A.R. Urquhart, "Cellulose-derivative Rayons", J. Text. Inst., 1951, 42, 385.
12. J.C. Arthur, Jr. J. Appl. Polym. Symp. 1989, 36, 201.
13. J.M. Cardamone, ACS Symposium Series, "Historic Textile and Paper Materials II", No. 410, S.H. Zeronian and H.L. Needles, Eds. Chapter 5, 1989, American Chemical Society, 231.
14. R.L. and J. Bogaard. Advances In Chemistry Series No. 212, H.L. Needles and S.H. Zeronian, Eds. Chapter 18, American Chemical Society, Washington, D.C., 1986, 330.
15. J.M. Cardamone, Ibid, 1988.
16. F. Shafizadeh and A.G.W. Bradbury, J. Appl. Polym. Sci., 1979, 23, 1431.
17. V.M. Skachkov and V.I. Sharkov, Izv. Vyssh. Ucheb. Zaved. Les. Zh., 1967, 10, 17. (CA 68, 96906C)
18. R.F. Nickerson, Ind. Eng. Chem., 1941, 33, 1022.
19. Herman, P.H. and A. Weidinger, J. Polym. Sci., 1949, 4, 135.
20. R.R. Muhhergee and T. Radhakrishnan, Text. Progress (Text. Inst.), 1972, 4, 4.
21. G.A. Serad and J.R. Sanders, "Encyclopedia of Tex-

tiles, Fibers, and Nonwoven Fabrics", M. Grayson, Ed.,
Wiley and Sons, N.Y. 1984, 65.
22. V. Daniels, In Preprints of the Contributions to the
Washington Congress, "Science and Technology In the
Service of Conservation", 1982, The International Institute for Conservation of Historic and Artistic Works,
London, 66.
23. D. N.-S. Hon, Idem., 92.
24. I. Block, Idem., 96.
25. N. Kerr, S.P. Hersh, P.A. Tucker, Idem., 100.
26. J.M. Cardamone and P. Brown. Advances In Chemistry
Series No. 212, "Historic Textile and Paper Materials",
H.L. Needles and S.H. Zeronian, Eds. Chapter 3, 1986, 41.
27. S. Hackney and G. Hedley, In Preprints ICOM Committee
for Conservation 7th Triennial Meeting, Copenhagen, 1984,
The International Council of Museums, 84.2.16.
28. G. Thomson, "The Museum Environment", 1978,
Butterfield & Co., London, 43.
29. J.H. Saunders, "Encyclopedia of Textiles, Fibers,
and Nonwoven Fabrics", M. Grayson, Ed., Wiley & Sons, New
York, 1984, 355.
30. M. Joseph, "Textile Science", 5th Ed., Holt,
Rinehardt and Winston, N.Y., 1987, 25.
31. N. Hollen, J. Saddler, A.L. Langford, S.J. Kadolph,
"Textiles", 6th Ed. Macmillan, N.Y., 1988, 11.
32. G.A. Serad and J.R. Sanders. Ibid., 68.
33. R. Meredith, J. Text. Inst., 1945, 36, T107.
34. R. Meredith, J. Text. Inst., 1945, 36, T147

The Degradation of Cellulose Triacetate Studied by Nuclear Magnetic Resonance Spectroscopy and Molecular Modelling

M. Derham, M. Edge, D. A. R. Williams, and D. M. Williamson

DEPARTMENT OF CHEMISTRY, FACULTY OF SCIENCE AND ENGINEERING, MANCHESTER POLYTECHNIC, CHESTER STREET, MANCHESTER M1 5GD, UK

Introduction

Nuclear magnetic resonance (nmr) spectroscopy is one of the most powerful analytical techniques available to chemists to study the structure of organic molecules from simple chemicals to bio-organic polymers and, indeed, polymers in general. The development of high field superconducting magnets, sophisticated electronics that allow reliable pulse sequences to be applied, and fast data acquisition, computing and display have combined to give a technique capable of detecting chemical species in the millimolar range of concentrations.

In the last decade the cost of computing hardware has declined sharply. This together with the design of graphical input and output software, and the transfer of molecular mechanics algorithms from mainframe computers to workstations has led to the development of molecular graphics and modelling. This technique which is becoming a routine tool for the organic and biochemist can usefully be applied to polymers of all types. In conjunction with nmr spectroscopy molecular modelling can provide detailed insights into macromolecular structures as well as the conformations and dynamics of small molecules.

This paper records some of our preliminary work using these two techniques on the degradation of cellulose triacetate, a polymer used extensively for such things as film stock, spectacle frames, and as an artistic medium.

Basic NMR Theory

There are many reviews and books about nmr spectroscopy both at a simple level[1] and with a higher theoretical content[2]. However, for those readers unfamiliar with this technique a short explanation seems appropriate.

A small sample of polymer (50 milligrams) is dissolved in a suitable solvent, eg deuterated chloroform, and placed in a very intense magnetic field. The spectrometer used for this work has a field strength of 6.3 Tesla, some 63000 times as strong as the Earth's magnetic field. The nuclei of the atoms of the molecules making up the polymer structure become magnetised in this very strong external field. They act rather like compass needles do and align themselves with the magnetic field. A short but intense magnetic pulse is applied to the sample using electromagnetic radiation in the radiofrequency range (270 MHz in this case). The nuclei can reorientate themselves in the magnetic field moving to higher energy levels (Figure 1) and these changes in energy correspond to very precise frequencies of the radiofrequency radiation. Thereafter, the nuclei relax back to their original lower energy state thus causing the change in magnetisation induced by the magnetic pulse to decay exponetially (Figure 2). This decay curve is called a Free Induction Decay or FID, and it contains all the information about what precise frequencies have been absorbed. However, this time domain spectrum has to be mathmatically changed or transformed into a frequency spectrum before useful information can be readily extracted. This change can be made using a mathematical tool called Fourier Transformation, but it involves a great deal of calculation and fast computers must be used. Once transformed the nmr spectrum can be displayed and plotted out.

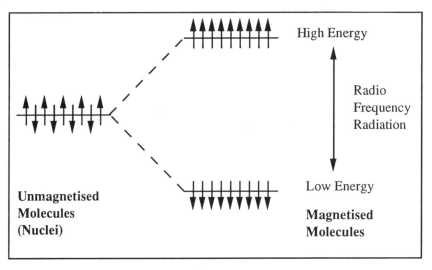

Figure 1
Energetics of NMR spectroscopy

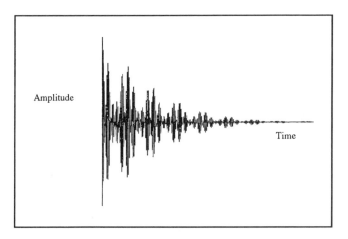

Figure 2
The Free Induction Decay for an Ethyl Group.

The description of nuclear magnetic resonance given so far can be likened to the sounding of a bell. Once hit by a clapper a bell will resonate, the sound being composed of many frequencies, and dying away with time. If we were to capture the sound on tape and feed that information into a computer we could carry out a Fourier Transformation and obtain a picture or spectrum of the frequencies composing the sound of the bell.

Figure 3 outlines the procedure for obtaining an nmr spectrum by this type of pulse or Fourier Transform (FT) nmr spectroscopy, while Figure 4 which shows the proton nmr spectrum for diethyl phthalate, a common plastisicer. The spectrum is said to be a proton nmr spectrum because it comes from the nuclei of the hydrogen atoms in the molecules of diethyl phthalate. The nuclei of hydrogen atoms are protons. Not all types of nuclei can resonate. In fact, only those comprising an odd number of protons and/or neutrons can give nmr spectra. Thus, the abundant isotope of carbon, ^{12}C, does not show nmr activity, but the less abundant isotope, ^{13}C, does exhibit activity and so we can also record so called 'carbon-13 spectra'.

The information in an nmr spectrum is characterised by four features. Refering to Figure 4, a number of peaks can be seen at various positions along the chemical shift scale. These peaks correspond to different frequencies of resonance and this reflects the kind of chemical or molecular environment in which the hydrogen nuclei find themselves. The methyl hydrogens (the CH_3 protons) are in a different environment from those attached to an aromatic ring, thus, they have a different resonance position. The changes in position are very small - of the order of parts per million, ppm, of the radiofrequency radiation. Thus, chemical shifts are referred to as so many ppm away from a standard reference - usually tetramethylsilane which is set to 0 ppm. Here the methyl resonances are at 1.5 ppm while the aromatics are at 7.6 and 7.8 ppm, showing that there are two types of aromatic proton.

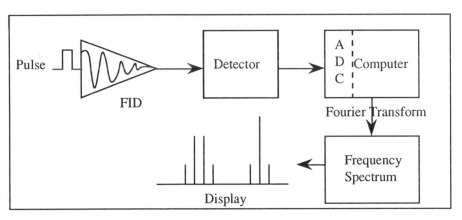

Figure 3
Basic NMR spectroscopy

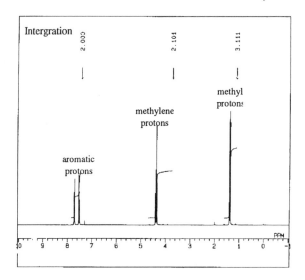

Figure 4
^{1}H spectrum of Diethyl Phthalate

The second feature that can provide information from the spectrum is the area under the peaks. For proton spectra the area under any peak is proportional to the number of protons giving rise to that peak. Thus, if the peaks are integrated the resulting step curves give information about how many protons of each type are present in the sample. These integrations are the primary source of quantitative data in nmr spectroscopy.

The third feature from which information can be obtained is the line width of the resonance peaks. In the case of diethyl phthalate all the peaks are very sharp. This is a reflection of the molecules moving about rapidly in solution and taking sometime, perhaps a second or more, to relax. Figure 5 shows the proton nmr spectrum of cellulose triacetate. By comparison to that of diethyl phthalate some of the peaks are very broad. This indicates that the polymer protons have very short relaxation times (< 0.1 s) and reflects the polymer molecule's slow movements in solution. Thus peak width relates to relaxtion times which in turn reflect how rapidly nuclei are moving in solution.

The final easily observed feature in nmr spectroscopy is the spin coupling pattern. Careful inspection of the methyl resonances in the spectrum of diethyl phthalate shows that there is a triplet structure, while those for the methylenes (the CH_2 protons) show a quartet. In both cases the patterns are characterised by specific ratios for the peaks (1:2:1 for the triplet and 1:3:3:1 for the quartet). These patterns can give (but not always) information about the number of protons on adjacent carbon atoms. However, spin coupling patterns are probably the least useful feature in polymer spectra and will not be considered further.

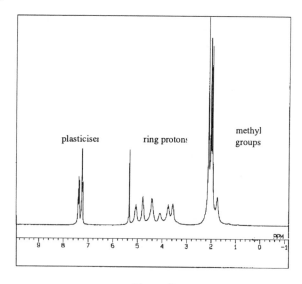

Figure 5
^{1}H Spectrum of Cellulose Triacetate

Cellulose Triacetate

The structure of cellulose triacetate (Figure 6) shows that it comprises a glycopyranose ring acetylated at C_2, C_3, and C_6, with β-1,4- linkages along the polymer chain. Although referred to as a triacetate not every position is acetylated and on average there are about 2.7 acetyl units per ring. The configurations of the substituents and the 1, 4 linkages are such that all the groups are equatorial. The macromolecular structure of the polymer chain is discussed later.

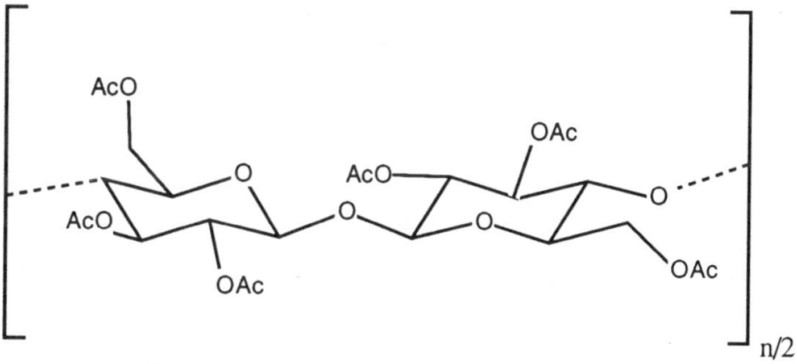

Ac = CH$_3$CO

Figure 6
Structure of Cellulose Triacetate Showing Arrangement of Cellulose Subunits.

Figure 5 shows the proton nmr spectrum of cellulose triacetate and gives the assignments of the resonances[3,4]. Note in particular that there are three acetyl resonances visible and that these are much sharper than the ring proton resonances. This would suggest that the acetyl groups are fairly free to rotate on the polymer chain which itself is much slower moving.

Some Examples of Degragation of Cellulose Triacetate

The first example of the use of nmr spectroscopy for looking at possible degradation in cellulose triacetate concerns a statue constructed from cellulose triacetate, in the 1930's. In order to assess whether any damage was occurring six one milligram samples were submitted for analysis. Figure 7 shows the nmr spectrum of one of those samples. The chemical shifts of the peaks show that this material has two plasticisers - diethyl phthalate and triphenyl phosphate. Integration of the appropriate peaks shows that these plasticisers are there in approximately equal amounts and that the total plasticiser content is about 20% by weight of the sample. By careful integration of the ring proton resonances and the acetyl resonances and correcting for peaks from plasticisers the amount of acetate present can be determined. In this case, these calculations suggest a value of 2.7 acetate groups per ring, in accord with normal cellulose triacetate. Thus, there is no clear evidence of any decay in this spectrum, but there is a suggestion from two small peaks at 1.0 and 1.2 ppm that there is some grease present. Note also that the solvent used, trifluoroacetic acid, gives rise to some impurity peaks.

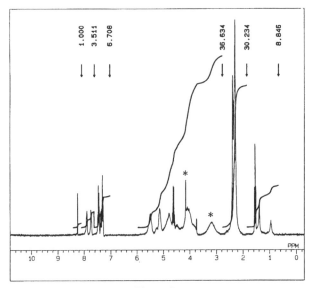

Figure 7
^{1}H Spectrum of Cellulose Triacetate in a Moulded State
(* - impurities from the solvent.)

In contrast, Figure 8 shows the nmr spectrum of a severely degraded sample of cellulose triacetate film stock. The acetate resonances around 2 ppm have disappeared and been replaced by a series of sharp peaks between 2.2 and 2.6 ppm. The ring proton resonances have changed in position and pattern. Also one of the plasticisers, diethyl phthalate, has undergone changes. The aromatic resonances at 7.8 ppm are degraded and a new triplet and quartet have appeared at 1.2 and 3.1 ppm respectively. These latter peaks strongly suggest that diethyl phthalate has decomposed hydrolytically to give ethyl phthalate and the ethanol so produced has reacted further with acetate to give ethyl acetate.

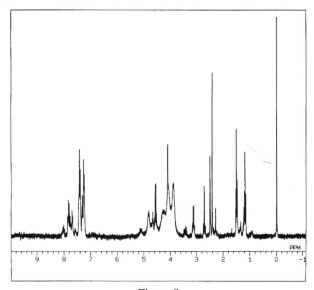

Figure 8
[1]H Spectrum of Severely Degraded Cellulose Triacetate Film Stock

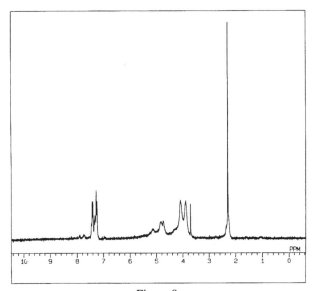

Figure 9
[1]H Spectrum of Partially Degraded Cellulose Triacetate Film Stock

Not all samples show such dramatic degradation. Figure 9 shows the spectrum of a more typical example. The acetate resonances at 2.2 ppm have changed, the two outer singlets having diminished in intensity, but a sharp singlet remains at 2.24 ppm.

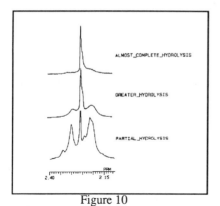

Figure 10
^{1}H Spectra of the Acetate Region in
Samples of Degraded Cellulose Triacetate.

The ring resonances have broadened even more and a sharp singlet at 3.6 ppm has appeared. Figure 10 shows the acetate region in more detail for three similar samples. Clearly, there are different levels of acetate loss in each case. From the assignment of cellulose triacetate noted earlier, the resonances that disappear are those from the C_6 and C_2 acetates. If a hydrolytic mechanism is involved then these observations are in accord with the known relative reactivity of the three different ring sites.[5] Figure 11 summarises the likely order of loss of acetate groups.

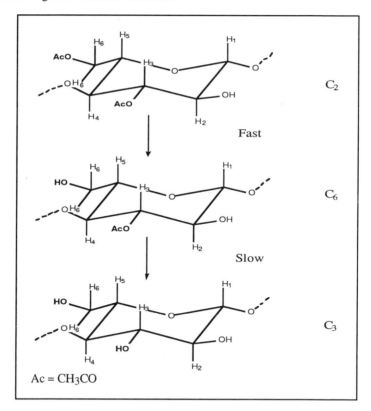

Figure 11
Probable Sequence of Acetate Loss in Cellulose Triacetate

With the previous examples the broad lines of the polymer resonances have been present. However, there is another type of nmr experiment known as a spin echo experiment that causes broad peaks to disappear from the spectrum leaving only narrow or sharp resonances. This type of experiment is useful for looking at partially obscured or totally hidden sharp lines.

The basis of the experiment is the difference in relaxation time between fast moving groups or molecules and the slower moving polymer chains. If the experiment is arranged so that the instrument does not collect the FID until some time has elapsed then the faster relaxing polymer protons will have decayed while the smaller faster molecules with longer relaxation times will still be decaying, ie will still give rise to a detectable signal. In order to detect this signal the spectrometer picks up an echo of the original FID. The process is also known as spin filtering and involves a multi-pulse sequence.

Figure 12a shows part of the normal nmr spectrum from a particular sample of cellulose triacetate which has partially degraded. At first sight there seems to be little different with this spectrum compared to the standard one (Figure 5); however, closer inspection of the ring proton region shows some sharper resonances underneath the broader peaks. Figure 12b shows the result of a spin echo experiment on this sample. Clearly, the broad resonances have all but disappeared leaving the sharp lines exposed. Analysis of the precise position and coupling pattern of these lines strongly suggests that they arise from the alpha and beta anomers of end groups on the polymer chains (Figure 13). Thus this particular sample appears to be undergoing a different type of degradation - some form of chain scission such that the number of end groups increases and become visible on the nmr spectrum.

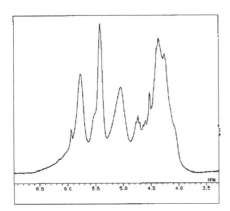 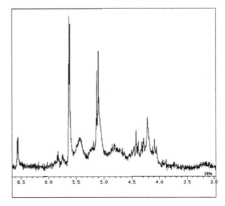

Figure 12a   Figure 12b
Normal Spectrum   Spin Echo Experiment
[1]H Spectra of the Ring Protons of Cellulose Triacetate in a Partially Degraded Sample

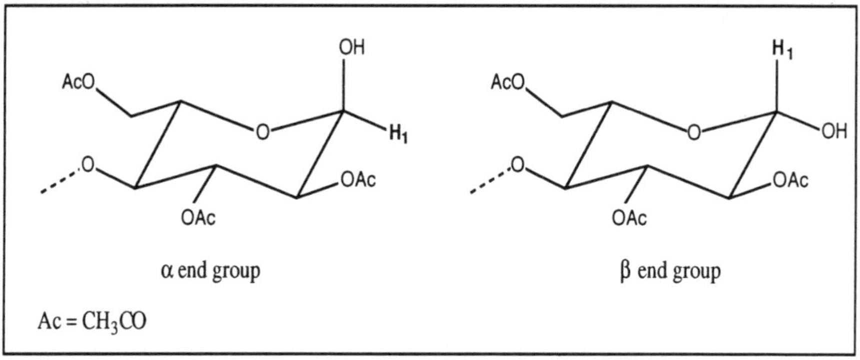

OH

AcO

O

H₁

OAc

OAc

α end group

H₁

AcO

O

OH

OAc

OAc

β end group

Ac = CH₃CO

Figure 13
α and β End Groups

¹³Carbon NMR Spectroscopy

As well as protons many other nuclei can exhibit nuclear magnetic resonance. According to the criterion laid out above carbon nuclei of isotope 13 can show nmr activity. As this isotope is only present in natural carbon containing samples at about 1.1% ¹³C nmr spectroscopy requires a considerable number of scans to be made before an adequate signal to noise ratio is attained. Often this means several thousands of scans taking many hours of experiment time. Also, the relatively insensitive nature of the ¹³C nmr spectroscopy means that the small amounts of products from degradation will be difficult to detect. However, the resulting spectrum shows a much enhanced chemical shift dispersion leading to the possible identification of signals and hence structures that might be hidden in the proton spectrum. Figure 14 shows the ¹³C nmr of cellulose triacetate and its assignment [6]. Note also that the plasticiser triphenyl phosphate is also present.

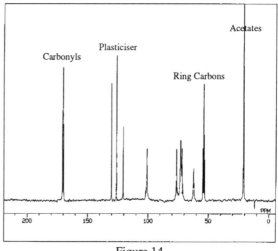

Figure 14
¹³C Spectrum of Cellulose Triacetate

Molecular Graphics and Modelling

While nmr spectroscopy can give great deal of information about the detailed molecular structure of a sample it cannot easily give much information about the macromolecular structure of regular polymers such as cellulose triacetate. To gain an understanding of what features may be present at a higher level the techniques of molecular graphics and modelling may be applied.

Figure 15 shows a comparison of the minimised structures of oligimers 20 units long of starch, cellulose, and cellulose triacetate. These structures have been obtained by drawing the monomer structures into a powerful molecular modelling package, QUANTA marketed and developed by Polygen. The monomers are then linked together to give any desired chain length and then a minimisation algorithm is applied to the polymer structure in order to find the minimum energy conformation of the structure. From these structures inter-atomic distances, bond angles, torsion angles, and so on can be measured with a reasonable degree of accuracy. The energy differences of different conformations can also be assessed.

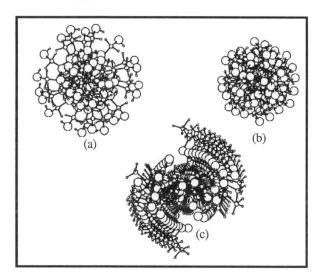

Figure 15
Minimised Structures for a)Cellulose, b)Starch and
c) Cellulose Triacetate, viewed along the Chain Axis

In this particular case, the conformations which are viewed along the inter-chain axis can be seen to be quite different. Cellulose and starch which differ by having a β-1,4- and an α-1,4- linkage respectively both show a high degree of helical symmetry along the chain axis. Cellulose triacetate on the other hand while showing helical symmetry displays a much opened helical coil. This change can only be ascribed to the steric or crowding effect of the three acetate groups forcing the coil into this alternative arrangement. Clearly visible are two sets of acetyl groups which upon closer examination turn out to be the C_6 and C_2 acetyls. Thus the two most reactive sites as seen by nmr spectroscopy are the two most exposed sites in the molecular model - a pleasing agreement.

The structure of cellulose triacetate shown in Figure 15 is a fully acetylated one. However, the degree of acetylation is, in fact, only about 87%. Figure 16 compares the fully acetylated structure to that of a model of cellulose triacetate with only 87% of its sites acetylated. The acetylation has been randomly generated by constructing monomers containing zero, one, two, and three acetyl groups respectively and combining these in a manner such as to give overall 87% acetylation. Clearly, the general nature of the macromolecular structure remains the same while some degree of symmetry has been lost.

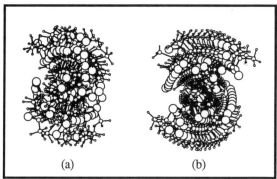

(a)   (b)

Figure 16
Minimised Structures for a) Partially and b) Fully Acetylated Cellulose

A number of questions can be posed about this macromolecular structure and its relationship to degradative pathways. Firstly, if the principal degradation mechanism is deacetylation, then does this occur in a random manner along the chain or does it start at a site and then work outwards from that site via a neighbouring group effect? Knowing which pathway is followed is important in the design of a stabiliser for such systems. If a neighbouring group mechanism is involved then a substance that could chelate several spatially close hydroxyl groups may be indicated. If a random mechanism is involved a more general type of stabiliser may be needed.

Secondly, what is the role of the plasticiser? How does it bind into the polymer structure and how does it alter the conformation? Using the analogy of drug-receptor theory can a better plasticiser be designed that could both stabilise and 'plasticise' the polymer?

Thirdly, the structures shown in Figures 15 and 16 are static structures. What would occur if these structures were given some energy to move around? How would the conformations alter if a dynamic system were constructed?

Fourthly, what is the effect of a supramolecular structure upon the energy and conformation of cellulose triacetate. A supramolecular structure could involve a double or even triple strand of polymer[7].

All these points are potentially answerable through molecular modelling, and our studies continue in these directions.

Conclusions

Nmr spectroscopy can be used effectively on small samples to characterise and identify qualitatively and quantitatively the components of polymer materials.

In the case of cellulose triacetate information about degradative pathways can be found - particularly about deacetylation mechanisms.

Use of more sophisticated nmr experiments such as the spin echo procedure gives further insight into the character of the polymer and any freely moving small molecules or parts of the polymer chain. In particular with cellulose triacetate there is some evidence of chain scission resulting in the production of alpha and beta anomeric end groups.

Molecular graphics and modelling gives an important insight into not only the detailed molecular structure of the monomer units comprising cellulose triacetate, but also into the macromolecular structure of the polymer chain. There is a strong suggestion that cellulose triacetate has a significantly different macromolecular structure from cellulose itself and this may have a bearing on the mechanisms of degradation.

Experimental

All nmr spectra were run as about 5% solutions in deuterated trifluoroacetic acid or deuterated dichloromethane both obtainable from Aldrich Chemical Co. Spectra were obtained on a JEOL 270 MHz GSX nmr spectrometer. Typical acquistion conditions for proton spectra were 8 scans of 16K data points over 5400 Hz, an acquisition time of 1.37 seconds using a pulse angle of 30°, with a pulse delay of 2 seconds. ^{13}C spectra were obtained with 16,000 scans of 16K data points over 5400 Hz, an acquisition time of 0.41 seconds using a pulse angle of 60°, with a pulse delay of 2 seconds.

Molecular modelling studies were carried out using QUANTA software by POLYGEN Corporation, running on a Silicon Graphics Personal Iris 40D computer.

Acknowledgements

The authors are grateful for technical assistance from Mrs C Cameron, Mr R P Warren, Mr J A Russell, and for financial aid from the Tate Gallery, London, UK.

References

1) D. A. R. Williams, 'NMR Spectroscopy', Analytical Chemistry in Open Learning Series, John Wiley & Sons, London, 1986.

2) A. Derome, 'Modern NMR Techniques for Chemistry Reasearch', Pergamon Press, Oxford,1987.

3) D. Gagnaire, F. R. Taravel, M. R. Vignon, Macromolecules, 1982, 15, 126.

4) B. Casu, Chapter 1, 'Polysaccharides: Topics, Structure and Morphology', E. D. T. Atkins (ed), Macmillan, 1985.

5) V. W. Goodlett, J. T. Dougherty, H. W. Patton, J Polym Sci, A-1, 1971, 9, 155.

6) T. Miyamoto et al, J. Polym. Sci., Polym. Chem. Ed., 1984, 22, 2363.

7) C. M. Buchanan, J A Hyatt, D. W. Lowman, J. Am. Chem. Soc., 1989, 111, 7312.

Investigation of the Archival Stability of Cellulose Triacetate Film: The Effect of Additives to CTA Support

Y. Shinagawa, M. Murayama and Y. Sakaino

ASHIGARA RESEARCH LABORATORIES, FUJI PHOTO FILM CO. LTD, MINAMIASHIGARA, KANAGAWAKEN, 250-01, JAPAN

1 INTRODUCTION

It is well-known that cellulose tri-acetate (CTA) and poly ethylene terephthalate (PET) are predominantly used as photographic supports. The chemical structures of the supports are shown in Fig.1. Recently it has been shown that CTA film support degrades at certain storage conditions[1-6]. The physical and chemical properties of PET stored at various unfavourable storage conditions are far superior to those of CTA stored at normal conditions[7-11]. However CTA support has a lot of advantages over PET support in practical handlings. For instance, it can be conveniently spliced by a film cement, and easily rejuvenated by organic solvent[12]. Another advantage of CTA film support lies in its easier recovering from the core set curl compared to PET. We studied the archival stability of both of the two photographic support and found that the instability problem related to archival storage is more serious in CTA support. This paper reports our studies on CTA base stability.

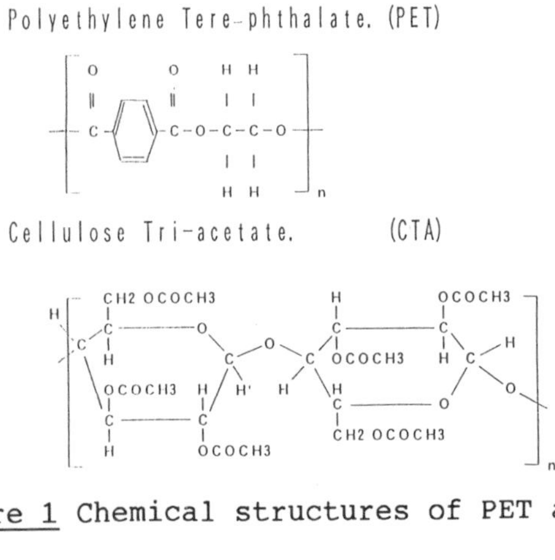

Figure 1 Chemical structures of PET and CTA

As was mentioned by T.Ram[10], CTA film is stable at
the optimum storage conditions specified in ANSI[13].
However, it is fairly difficult to keep CTA film in the
condition recommended for long time storage. Our studies
consequently focused on how we can make CTA more stable
under non-ideal storage conditions. Our target was to
develop the method for improving CTA support stability from
the view point of film manufacturing.

2 EXPERIMENTAL

A schematic illustration of photographic film is shown in
Fig.2. A number of ingredients incorporated in the
support or in the coated layers thereon must have more or
less effects on the archival stability of the film. The
influences of some ingredients on the archival stability
have already been reported[9].We carried out similar
experiments with a special stress on the additives to CTA
support.

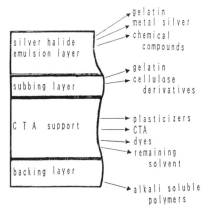

Figure 2 Schematic cross-sectional view of photographic film

    Materials. CTA is produced by acetylation of the
alcohlic moiety of cellulose. Its scheme is shown in Fig.3.
Cellulose has three hydroxyl groups in a glycosidic ring.
Fig.3 shows the case for the degree of acetyl substitution
being 3 where all the hydroxyl groups are completely
acetylated. The degree of acetyl substitution is 3. The
following experiments were performed with CTA flakes usually
used in the manufacturing line of Fuji Photo Film Co.; they
have the degree of acetylation of 2.9 and 310 recurring
units in one molecule.

    Preparation of Support Samples. CTA flakes were
dissolved in a solvent mixture comprising methylene chloride,
methanol and n-butanol to form a high viscosity CTA solution
called "dope". Plasticizers and stabilizers were added to
the dope if necessary for the purposes of experiment. As is
shown in Fig.4, the dope was cast on a slowly rotated

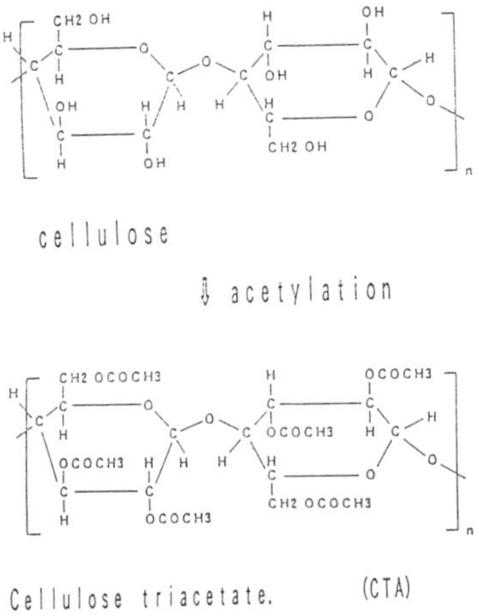

cellulose

↓ acetylation

Cellulose triacetate.   (CTA)

Figure 3 Preparation of CTA from cellulose

stainless endless belt and then stripped off therefrom. The
stripped CTA support was then dried through the dryer zone,
and taken up to form a roll. The amount of remaining solvent
was carefully contolled by adjusting the drying conditions.
Then subbing layers and emulsion layers were coated when
required for testing. The CTA support thus prepared is
typically specified as;
Triphenyl phosphate (TPP) : 16% of CTA weight,
Remaining n-butanol   : 1% of total support weight
Support thickness   : 140 μ m.

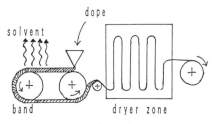

Figure 4 Diagram of CTA support production

    Accelerated Aging Test.   The support samples were
degraded at various accelerated conditions according to
the following procedure: Two grammes of CTA film specimens
were put in a 15 ml volume glass container and kept at 90℃ ,
100%RH (Relative Humidity) for 24 hours for preconditioning
under unsealed condition. At the end of preconditioning, the
container was sealed and kept at the same condition. When
plasticizer powder was tested, 2 g of plasticizer was put in
the glass container and was subjected to the same procedure.
Fig.5 illustrates this method schematically.

    Naturally Aged Samples.   Many naturally aged films
were submitted from our users. Among those, two samples of
B/W negative films with different aging histories were
analyzed. One of them had been stored at a user's house

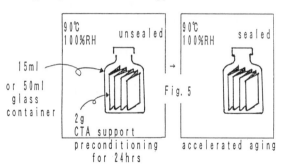

<u>Figure 5</u> Accelerated aging test

naturally, sealed in an iron can and the atmospheric
condition depended on the climate. The storage condition may
have drifted between 35℃ 100%RH and 0℃ 10%RH,
considering the climate of Japan. The other had been
preserved constantly at 23℃ 55%RH for 33 years in a sealed
tin plated iron can.

<u>Sample Preparation for Analysis.</u> Before analysis,
each photographic film sample was soaked in a 1 wt% aqueous
solution of a protein-digesting enzyme (Bioplase SP-4,
Nagase) at 40℃ for 1 hr, then rinsed to remove emulsion
layers. Thus, support without coatings was obtained. The
backing and subbing layers were scraped by a razor. Then,
the plasticizers were removed out of each support sample
by rinsing four times in hot methanol at 40℃.

<u>Viscosity Measurement.</u> A decrease of molecular weight
was detected by viscosity measurement in trifluoro-acetic
acid (TFA) solution. The emulsion layer, subbing layer and
plasticizers were removed from specimens before dissolving
specimens in TFA solution. The solution was kept below 5℃
until viscosity measurement to prevent degradation of CTA in
the solution. The mixed solvent of methylene chloride and
methanol was unsuitable for the viscosity measurement, since
the resulting solution failed to give a sufficient level of
reproducibility of measurement. One reason lies in a poor
dissolving capability for degraded CTA manifested by a
significant amount of insoluble residues left. The flow rate
of 1 w/v% TFA solution of each sample was measured with by
Ubbelhode type viscometer (The flow time of water at 30 ℃
was 30 sec). Viscosity retention (%) was calculated by the
following formula (1),

$$\text{Viscosity retention (\%)} = \frac{t_{aged} - t_{solvent}}{t_{fresh} - t_{solvent}} \times 100 \quad : (1)$$

wherein t_{aged} is the flow time of the aged sample solution,
t_{fresh} is that of fresh sample solution, and
$t_{solvent}$ is that of the solvent, respectively.

Determination of Degree of Acetate Group Substitution.
^{1}H-NMR spectrum of each sample was obtained by Bruker WM400 from 0.4 w/v% trifluoroacetic acid-d (deuterated) solution. Using peak areas assigned as methyl protons of acetyl-methyl group and glycosidic ring protons, the degree of acetate group substitution was calculated.

Determination of Plasticizer Decomposition.
Plasticizers were extracted from each degraded CTA support by 40 v/v% aqueous acetonitrile solution for 1hr. The extractions were identified and quantified by reverse phase HPLC. The gradient condition of HPLC was as follows;
Eluent A: acetonitrile/water/1N HCl=300/1200/10(v/v/v),
Eluent B: acetonitrile/water/1N HCl=1200/300/10(v/v/v),
0-10min: A/B=66/34(v/v), constant,
10-15min: A/B=66/34(v/v) to 34/66(v/v), linear,
15-40min: A/B=34/66(v/v), constant.

Molecular Weight Measurement.
Degraded CTA samples usually cannot be dissolved in any ordinary solvents for Gel Permeation Chromatography (GPC). Thus, each sample was firstly peracetylated by trifluoroacetic anhydride following the method of Bourne et al.[14], and then perfectly dissolved in methylene chloride for molecular weight determination by GPC. After this procedure (i.e., 'Re-acetylation'), the re-acetylated CTA can be dissolved in methylene chloride completely and was subjected to GPC for molecular weight determination. We confirmed that the reacetylation process did not induce reduction of molecular weight. The condition of GPC was as follows;
Column: Shodex K803,805,806 in series (Showa Denko)
Eluent: Methylene chloride
Detection: Refractive Index Detector (RI)
Calibration: Polystyrene standards, Mw;5400-4880000

3 RESULTS AND DISCUSSION

Effect of TPP on CTA degradation.
Tri-phenyl phosphate (TPP) is a most popular plasticizer of photographic CTA supports. The effect of TPP content on viscosity retention of CTA at an accelerated aging (90°C ,100%RH) is shown in Fig. 6. Viscosity does not reflect accurately the decrease of molecular weight because it also depends on the degree of substitution of acetate group. Still it is a convenient method to detect the overall degradation. Fig.7 shows the relationship between viscosity retention and molecular weight, wherein the viscosity retention was measured with degraded CTA, but molecular weight was obtained by GPC measurement of re-acetylated samples. Since a good correlation exists between molecular weight and viscosity retention, it is confirmed that increasing TPP content induces rapid degradation of CTA. Furthermore, it is important that the plasticizer-free CTA support is much more stable than the samples plasticized with 10.7% or 16.3% TPP.

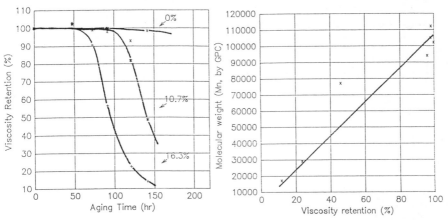

Fig.6 The effect of TPP content on Viscosity Retention of CTA at an accelerated condition (90C 100%RH)

Fig.7 Molecular weight versus Viscosity Retention

The Effect of Various Plasticizers. Fig.8 and Fig.9 show viscosity retention for several kinds of plasticizers. It can be seen that the extent of viscosity retention depends on the type of plasticizer. Phosphate plasticizers caused a significant degradation, but phthalate plasticizers little affected CTA degradation. HDDPP initiated the degradation at a very early stage of accelerated aging. It seems that the behaviour of CTA support degradation depends on the decomposing properties of the individual plasticizers.

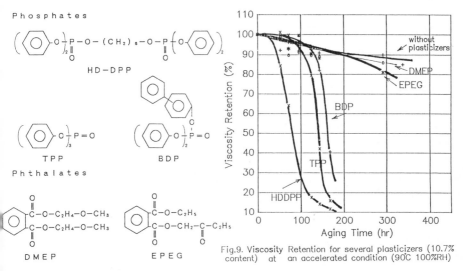

Fig.9. Viscosity Retention for several plasticizers (10.7% content) at an accelerated condition (90C 100%RH)

Fig. 8 Chemical structures of several plasticizers for CTA support

TPP : Tri−phenyl phosphate
BDP : Bi−phenyl di−phenyl phosphate
HD−DPP: 1,6−hexanediol−bis (di−phenyl phosphate)
DMEP : Dimethoxyethyl phthalate
EPEG : Ethyl−phthalyl ethyl glycolate

     Hydrolysis of TPP. In order to clarify the mechanism
of the decreasing viscosity, the following experiments were
conducted using TPP flakes. In 100 ml water immersed 1
gramme of incubated TPP flakes to extract the acid
ingredients generated by the decomposition. Two incubation
conditions were chosen; 90℃ and dry, and 90℃ and 100%RH.
The results are shown in Fig.10. pH change was not observed
at the dry condition, while in the wet condition, two stage
pH decrease was observed; An initial slow pH shift followed
by a rapid decrease, taking place approximately over 50hr.
This means that TPP is easily hydrolysed and generates acids
at wet conditions though it is fairly stable at dry
conditions. Next, the decomposition behaviour of TPP was
examined.

     Quantitative Analysis of Acid. In an aqueous n-Butanol
solution was dissolved 1 gramme of incubated TPP at the
concentration was nine to one in v/v. The solution was
titrated by 0.01 N potassium hydroxide solution in ethanol.
In Fig.11, the ordinate represents acidity value, equal to
the weight of potassium hydroxide in KOH mg consumed per g
TPP. and the abscissa represents aging time at wet condition.
TPP decomposes and generates acids rapidly. The curve shape
in Fig.11 indicates an auto-catalytic decomposition over a
certain threshold of acid concentration. The calculated pH
from the acidity value assuming the complete dissociation of
generated acid coincides with the observed pH value fairly
well.

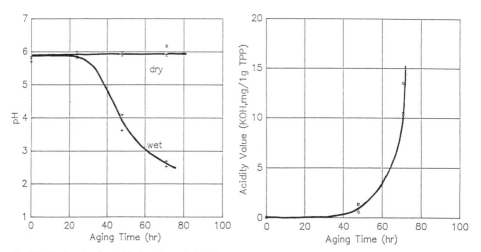

Fig.10 pH of water in which the degraded TPP are
       immersed versus accelerated aging time

       dry : 90℃ and dry condition
       wet : 90℃ and 100%RH

Fig.11 Acidity Value of the degraded TPP powder
       (KOH,mg/1g TPP)
       at an accelerated aging condition
       (90℃ 100%RH)

     From the above result, we decided to determine TPP
and its decomposed products in the heavily degraded CTA
support, using HPLC. The results shown in Fig. 12 indicate
that TPP is hydrolysed in the degraded CTA support sample.

Using standard materials, the four peaks were confirmed to correspond to MPP, DPP, phenol and TPP respectively. Fig.13 plots the molar amount of TPP and the decomposed products therefrom degraded CTA support samples. TPP decomposes to DPP and Phenol. The molar amount of decomposed TPP corresponds to those of generated DPP and Phenol over the entire range of aging time examined.

DPP is a strong acid. In fact 10mM aqueous solution of DPP showed pH 2.1, indicating a complete dissociation of the acid. The pH values estimated from the amount of generated DPP at 100 hr and longer are much lower than 2. This is based on the assumption that generated DPP dissolves in water contained in CTA support (the equilibrium content of water at 90 ℃ 100%RH is 10 wt% CTA support). But in reality the DPP is in partition equilibrium between the oleophilic bulk and water. It is strongly suggested that, apart from a precise mechanism, the hydrolysis of CTA proceeds rather rapidly when the concentration of DPP in the water contained in CTA exceeds a certain threshold value. The analytical results clearly show that TPP decomposes to DPP and phenol.

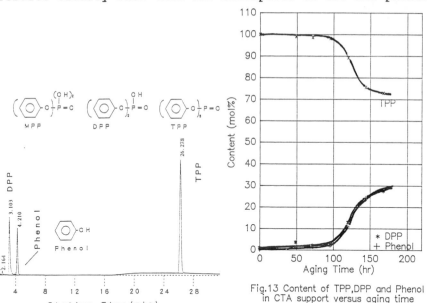

Fig.12 Liquid Chromatogram of TPP and its decomposed compounds extracted from CTA support degraded at 90℃ 100%RH for 168hrs.

Fig.13 Content of TPP,DPP and Phenol in CTA support versus aging time

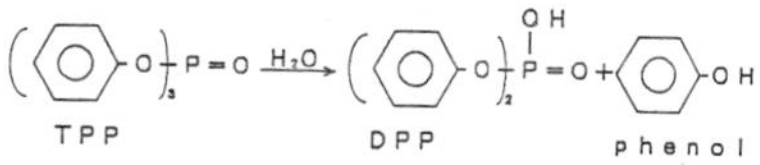

Relationship between CTA Degradation and TPP Decomposition. In the upper side of Fig.14, viscosity retention and DPP content are plotted versus accelerated aging time. At the early stage of aging viscosity is constant and does not show any decrease. Then a rapid decrease takes place at 100 hr. In advance of this rapid decrease, the content of DPP starts to increase though rather slowly. and at 100 hr point, a rapid increase takes

place. The period up to the 100hr point can be regarded as
an induction period for viscosity retention in spite of the
presence of small amounts of TPP.

    Along with the molecular weight reduction, some change
in the degree of substitution of acetate group must take
place: we measured the degree of substitution of acetate
group by ^{1}H-NMR and also determined the molecular weight by
GPC using re-acetylation procedure. The results shown in
lower side of Fig.14 indicates that the decrease in the
degree of substitution and the reduction of molecular weight
occur simultaneously.

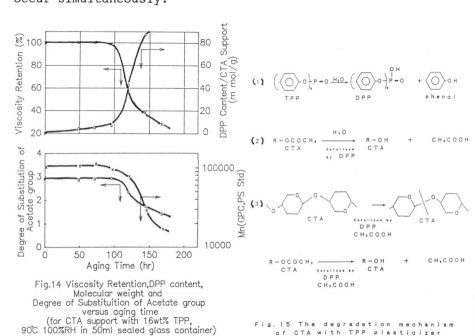

Fig.14 Viscosity Retention,DPP content,
Molecular weight and
Degree of Substituition of Acetate group
versus aging time
(for CTA support with 16wt% TPP,
90℃ 100%RH in 50ml sealed glass container)

Fig. 15 The degradation mechanism
of CTA with TPP plasticizer

    The results which were obtained from accelerated aging
CTA support samples will be summarized as;

    1. CTA support without TPP plasticizer is very stable
       even at wet conditions.
    2. TPP is stable only at dry conditions.
    3. TPP releases DPP at wet aging conditions.
    4. DPP is a strong acid and by the generation of DPP,
       the pH value of CTA support can decrease to as low
       as 2.0.

    As was mentioned by many researchers, it seems correct
that the degradation of CTA support is caused by the acetic
acid catalyzed autocatalytic hydrolysis. However, we also
think that DPP generated by the decomposition of TPP has a
more significant effect on CTA degradation. As the present
results suggests, DPP is the trigger to initiate the

hydrolysis reaction. The situation is shown in Fig.15. We should not ignore the effect of DPP. As was shown by Vos, et al.[15], the hydrolysis rate of cellulose ester is minimum at a pH value between 4 and 5.

How the CTA degradation proceeds with aging time is shown in Fig.16: a three stage degradation model is proposed based on our experimental results.

At the first stage a small amount of DPP is generated from TPP, but pH is not low enough to initiate the hydrolysis of CTA. At the second stage, pH decreases to around 3 and the degradation begins to proceed but with a slow rate. As the pH of the system approaches 2, the final stage of rapid hydrolytic degradation comes whereby both DPP and acetic acid will probably attack the CTA chain.

The details of the degradation mechanism were not clarified completely yet. In order to stop the degradation of CTA, we have to do a lot of study to understand the mechanism and to find out effective remedies. With reference to our present results, the stability of TPP-plasticized-CTA support would be much improved by preventing the decomposition of TPP to DPP or by deactivating the acid activity of DPP.

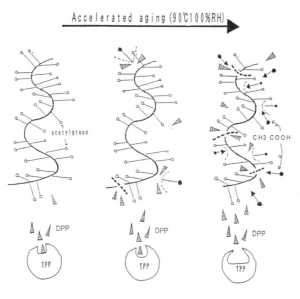

Accelerated aging (90℃100%RH)

Induction stage   Onset stage   Rapid degradation
I   II   stage III

Fig. 16
Hypothesis of CTA degradation by DPP

Naturally Aged Photographic Films. The analysis of
naturally aged photographic films was also carried out.
Emulsion layers and backing layers of sample films were
removed before testing. Degree of substitution, molecular
weight, DPP content are shown compared with accelerated
aging samples in Table 1.

Table 1. Specifications (and analytical results) for the
support films in the naturally aged photographic samples
with reference to those of the accelerated aging samples.

Samples	Degree of substitution by ¹H-NMR	Molecular weight by GPC (PS Std)	DPP content (μmol/g)	Duration of storage
Naturally aged samples stored at a user's house				
#166	1.78±0.07	73,000	32.8	33 years
B/W #172	1.90±0.02	78,000	31.1	33 years
negative #227	2.06±0.06	90,000	29.9	31 years
#237	1.62±0.02	61,000	22.9	31 years
#375	2.52±0.10	118,000	13.4	26 years
preserved at a library in Japan (23℃ 55%RH)				
#A1	2.92±0.05	128,000	5.0	30 years
B/W			(no Vinegar odor)	
negative #A2	2.88±0.02	128,000	8.6	30 years
			(Vinegar odor)	
Reference: accelerated aging samples (90℃ 100%RH)				at 90℃ 100%RH
#0	2.91±0.06	102,000	0.9	fresh
#1	2.85±0.17	105,000	4.0	49 hours
#2	2.89±0.12	112,000	5.0	72 hours
#5	2.92±0.04	94,000	6.9	96 hours
#8	2.33±0.04	77,000	42.7	120 hours
#9	1.73±0.05	29,000	87.5	144 hours
#12	1.40±0.01	17,000	105.6	168 hours

DPP was detected also in naturally aged samples.
Especially large quantities of DPP were detected in films
which were stored at a user's house. In case of a B/W
negative (#166) which was aged at a user's house for 36
years, the amount of DPP reached to 32.8 μ mol/ g CTA.
A considerable amount of TPP deposit was observed on the
film surface and strong odour of vinegar was noticed as
well. Since the level of 32.8 μ mol DPP corresponds to 115
hr of the accelerated aging, this sample is regarded as
well in the third stage of degradation.

In another case, two B/W fine grain negatives have
been preserved in a library in Japan. A constant
preservation at 23℃ and 55%RH which is thought close to

ideal still generated 5.0 and 8.6 μ mol DPP, respectively.
The sample containing 8.6 μ mol DPP, had the odour of
vinegar. Judging from the amount of DPP, it is concluded
that even those samples stored at such library conditions
are already at the onset stage towards rapid degradation.

The data in Table 1 were plotted in Fig.17. In the
figure, both molecular weight and the degree of acetyl
substitution decrease corresponding   to the increase of DPP.
Although the behaviour of naturally kept samples differs from
those of accelerated aging samples in detail, we think that
DPP behaves as the trigger of hydrolysis in this case, too.

Improvement of The Stability of CTA.   Three compounds
that were reported by Allen et al.[16] were used as
additives to a "dope". The "dope" with additives were cast
by the previously mentioned method. The three compounds are
sodium phenylphosphinate (SPP), a hindered amine Tinuvin 770
(T770) and a hindered phenol and chelate hybrid reagent
Naugard XL-1 (NXL-1). The results are shown in Fig.18. The
results with NXL-1 only is not shown in the figure because
it showed no effect of stability improvement. The CTA support
with the three compounds was the most stable as shown in this
figure. This result coincides with those of soaking
experiments reported by Allen et al.[16]   Once we know the
precise mechanism of CTA degradation, several effective
measures may well be adopted by photographic film
manufacturers.

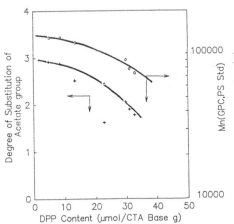

Fig.17. Relationship between Mn, the degree of
substitution of acetate group and
DPP content on naturally aged samples

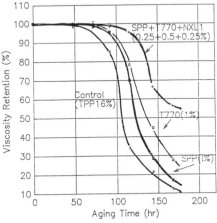

Fig.18 Improvement of CTA degradation by
additives. Viscosity retention versus aging time

4 CONCLUSIONS

__1.__ Unplasticized CTA film is very stable and does not degrade even at severe accelerated conditions.

__2.__ Our accelerated aging tests on TPP-plasticized photographic CTA support strongly suggests an important role of DPP generated by the decomposition of TPP in the deterioration of the support.

__3.__ A close correlation between DPP content and viscosity retention of CTA was observed both in artificially and in naturally aged photographic films. This fact indicates that the above assumption is applicable to the natural aging mechanism.

__4.__ Prevention of DPP generation or reduction of the DPP activity as acid would surely improve the archival stability of photographic CTA support.

REFERENCES

1. G. Pollakowski, Bild und Ton, 1987, 40, 90.
2. N.S. Allen, M. Edge, C.V. Horie, T.S. Jewitt and J.H. Appleyard, J.Photogr.Sci., 1988, 36, 103.
3. N.S. Allen, M. Edge, J.H. Appleyard, T.S. Jewitt, C.V. Horie and D. Francis, European Polymer J., 1988, 24, 707.
4. N.S. Allen, M. Edge, J.H. Appleyard, T.S. Jewitt, C.V. Horie and D. Francis, Polym.Degr.Stab., 1987, 19, 379.
5. M. Edge, 'The Degradation and Stabilization of Archival Cellulose-Ester Base Cinematographic Film', PhD Thesis, Manchester Polytechnic, Manchester, UK, 1988
6. N.S. Allen, J.H. Appleyard, M. Edge, D. Francis, C.V. Horie and T.S. Jewitt, J.Photogr.Sci., 1988, 36, 34.
7. P.Z. Adelstein and J.L. McCrea, Photogr.Sci.Eng., 1965, 9, 305.
8. P.Z. Adelstein and J.L. McCrea, J.Appl.Photogr.Eng., 1981, 7, 160.
9. A.T. Ram and J.L. McCrea, SMPTE J., 1988, 34, 474.
10. A.T. Ram, Polym.Degr.Stab., 1990, 29, 3.
11. A.T. Ram, S. Masaryk-Morris, D.Kopperl and R.W.Bauer, Imag.Tech., 1991, 124.
12. K. Brems, Imag.Tech., 1991, 94.
13. ANSI PH1.43-1985, Photography (Film), 'Processed Safety Film -Storage-', 1985.
14. E.J. Bourne, M.Stacey, J.C.Tatlow and J.M.Tedder, J. Chem.Soc., 1949, 2976.
15. K.D. Vos, F.O. Burris,Jr. and R.L. Riley, J.Appl.Polym. Sci., 1966, 10, 825.
16. N.S. Allen, M. Edge, T.S. Jewitt, and C.V. Horie, J. Photogr.Sci., 1990, 38, 26.

Film Storage

Morten Jacobsen

DANCAN INTERNATIONAL SALES, CARIT ETLARS VEJ 5, 1814 FREDERIKSBERG C, COPENHAGEN, DENMARK

Determining new Standards for long term Storage of Film

"The Vinegar Syndrome" is a new phrase in our vocabulary. The Manchester Polytechnic have for the past decade researched in degradation of polymers. They have worked on a project initiated by the National Film Archive in England to find solutions to the degradation of film.

This article will look at standards over a period of 30 years in long term storage of film and recommend new standard procedures in the light of research and findings.

1. Introduction

The Centre for Archival Polymeric Material at Manchester Polytechnic have in their opening remarks in their vast report from 1989 said the following: "Research is now needed into the causes, innate or external, of the degradation of materials used in the storage of archival media. These materials are unstable, subject to chemical reactions of oxidation, hydrolysis or photolysis, leading to loss of strength, yellowness, cracking or fading. Catastrophic failures of archival audio-visual materials are recent examples of failure in planning for the future.

We,separately, and as a group, have established reputations in conservation, polymer stabilising and image technology. These are necessary and complementary disciplines if the pressing problems are to be successfully tackled. A precise definition of the problem would require the following objectives to be gotten.

To identify the nature of degrading agencies present in a particular environment. To understand the nature of the physical and chemical effects that these agencies have on component structures and properties. To understand and quantify the interactions following combined agency attack. Our aim is to provide an expert consultancy service on two fronts:

1. To develop specifications for archival stable materials preferably lasting more than 200 years.

2. To develop methods to conserve existing materials used in archives and museums".[1]

2. International Standards or Recommended   Practice

I should like to give the reader a certain background in understanding the latest developments. I therefore think it is necessary to look at the different world standards or recommended practices before commenting.

2.1 SMPTE - PR131[2]
Storage of Motion-Picture Films
(5.3.2.) **Storage enclosures** (For full details see the recommendation)
The recommendations in this clause are taken from clause 3.1 of ANSI PH1.43-1985, and modified (Refer to that document for additional information):

"Storage enclosures for motion-picture films should preferably be steel reels and cans. Steel reels may be tinned, but not lacquered or laminated. Preferred plastic enclosure materials (cores and reels) are cellulose esters and polyolefins (polypropylene). Acceptable plastic materials are polystyrene and polyacrylates. Polyvinylchloride is not recommended for this use. Use of steel for reels is permissible provided that the reels are well protected by enamel, tinning, or other corrosion resistant finishes. Plastic that might give off reactive fumes or exudations during storage such as peroxides or chlorines, shall not be used".

The environmental conditions for humidity and temperature are as follows for archival storage conditions:

Silver-gelatin	Cellulose ester	15-50%	21°C
Silver-gelatin	Polyester	30-60%	21°C
Color	Cellulose ester	15-30%	2°C
Color	Polyester	15-30%	2°C

2.2 ANSI PH1.43-1985[3]
The American National Standard Institute is interpreted differently at some points and they should be mentioned. Interestingly enough under Medium-Term Storage Enclosures:

"Any film that releases acidic fumes or oxidants shall be stored in separate storage housings. Polystyrene or polyethylene containers are preferable to cardboard or metal containers for such films" and under Archival Storage Enclosures: "Containers should be of non-corrosive materials such as anodized aluminium, stainless steel, or peroxide free plastics".

2.3 EBU Tech. 3202-E 1974[4]
Page 65: The first stage in the long–term storage of films is thus an appropriate chemical processing. On the basis of the enquiry that has been analysed above, therefore the following summary of the procedures may be recommended for keeping programmes recorded on film in a good condition during storage for a sufficiently long time.

1.) Adequate fixing and washing
2.) Adequate drying
3.) Cleaning the boxes in which they are enclosed

4.) Placing the films within plastic bags inside the boxes
5.) Storage in rooms protected against flooding and industrial fumes, with the following ambient conditions
5.1.) Temperature 15°C+ 5–10°C
Care should be taken to avoid rapid changes from one temperature to a higher temperature (risk of condensation)
5.2.) Relative humidity 50% +/-10%
5.3.) Possibly, filtering the air admitted to the storeroom (a few microns)
6.) Storing the boxes horizontally

2.4 FIAF Preservation Commission
Preservation of Moving Images and Sound[5]
The Preservation Commission of the International Federation of Film Archives (FIAF) is an international working group that conducts research and publishes guidelines and recommendations on all aspects of moving image preservation.

The president is Dr. Henning Schou from the National Film and Sound Archive in Canberra, Australia.

The Preservation System II: Storage
(6.2) External factors affect the stability of materials.
The main factors affecting cinematographic film in dark storage are temperature, relative humidity and chemical contaminants in the atmosphere.
(6.2.1) Air purity
The air in many industrial areas contains small amounts of gases such as nitrogen oxides as from nitrate film, sulphur dioxide, ozone as from photocopying machines, and hydrogen sulphide. It is important therefore that a vault for archival storage of film be located where the air is clean, or else the air supplied to the storage area be purified and filtered for gases and dust.
(6.2.2.) Temperature and humidity
I will describe in detail the enormous significance of proper storage at low, steady temperature and relative humidity (RH) for the various types of material.

(6.4) Storage conditions for Black and White safety Film and Magnetic Tape
Recommended storage temperature
> Less than
> 16°C +/- 1°C on a daily basis. (+/- 2 on an annual basis)
> Relative humidity
> 35% +/-2% RH on a daily basis (+/-5 on an annual basis)
> Maximum range: 30-60% RH for film
>   20-50% for magnetic tape
Rate of fresh air intake is determined by national health regulations.

On storage of video tape, the Ampex Corporation recommends temperatures between 18° and 21°C and 20-40% RH. As stated above the environment should be as constant as possible, because fluctuations can cause damage through expansion and contraction of wound tape. Dust must be kept to an absolute minimum.

The Technical Committee of the International Association of Sound Archives
(IASA) has discussed the archival storage conditions for audio tape in great detail.
Do not store magnetic tape near stray magnetic fields.

Some archives seal safety film and magnetic tape in vapour-tight bags. But,
decomposing acetate film must not be sealed in bags from which the detrimental
decomposition gas cannot escape. (Vinegar syndrome)

(6.5) Storage conditions for colour film and diacetate film

Recommended storage temperature
-5°C +/- 1°C on a daily basis (+/- 2 on an annual basis)
Relative humidity
30% +/- 2% RH on a daily basis (+/- 5 on an annual basis)
Maximum range: 15-35% RH

Rate of fresh air intake is determined by national health regulations. Archivists
should not be disheartened by the low temperature recommended above, but always
keep in mind that every degree the temperature is lowered is beneficial. For example, a reduction of approximately 6°C will double the useful life of almost any
material.

Temperature and humidity effect colour dyes as can be seen from Tables (6.1)
and (6.2)

Table (6.1)

Affects of temperature on stability of image dye

Storage temperature	Relative storage time*
30°C	0,5
24°C	1
19°C	2
13°C	4
4°C	16
-18°C	340
-26°C	1000

Table (6.2)

Affect of humidity on stability of image dye

% Relative humidity	Relative storage time*
60	0,5
40	1
15	2

Generally, yellow is the dye most sensitive to humidity.

* Predicted time at 40% RH, for 10% loss of image dye. These values do
apply exactly to all dyestuffs, but they are close enough for most practical purposes.

Note the significant gain in longevity by reducing temperature and relative
humidity.

(6.8) **Storage in sealed bags**
You can greatly diminish the humidity problem by sealing stable safety film and magnetic tape in vapour-tight bags. Sealing takes place after the materials have accommodated themselves to approximately 20°C and 30-50% RH. In hot and humid countries this could be obtained more easily by using the film conditioning apparatus, also known as FICA, constructed by the Swedish Film Institute.

(6.8.1) **Diacetate in sealed bags**
Diacetate film may benefit from sealing in vapour-tight bags as this will reduce the evaporation of volatile plasticiser.

(6.8.2) *Warning* regarding use of bags
The wrapping of nitrate film or sealing in plastic bags is detrimental to the film. It is important for the decomposition gases to escape since it will otherwise accelerate the disintegration of the film.

Also, unless you are absolutely sure that the acetate film is stable (that is, no vinegar syndrome), the film should never be sealed in bags as stated above.

(6.9) **Storage rules**
(6.2.9) **Methods of storage.**
The storage method is also important, since in long term storage, various stresses and strains in the material become significant. The FIAF Preservation Commission recommend certain rules about the canning and storage of preservation copies of film and video tapes. These are some of these rules:

Store film cans horizontally in stacks preferably no higher than about 30 cm and, whenever possible, have all cans in the stack of the same diameter.

Store video tape upright on shelves, with winding axes horizontal. Do not stack video tapes on top of one another (8.4)

The containers should contain

1) the film and a core or spool only, or

2) video tape only - no paper, household plastic bags or other materials. In some circumstances, you may consider it desirable to enclose the film in special, vapour-tight bags. But, be aware of the vinegar syndrome.

Cans for nitrate films should have one or more holes in the side (preferably toward the bottom) to permit the escape of decomposition gases.

2.5 Author's Comments
This was 20 years of recommended practices with large variations.

The basic points to discuss are as follows:
1. Humidity
2. Temperature
3. Storage material

FIAF definately increased their demands to lower humidity. Storage of black & white acetate film is lowered from 60% to a dazzling 35% RH and the temperature increased from 6-12°C up to 16°C. When it comes to storage materials i.e. in cans, tins, storage enclosures or containers, then they do not recommend anything today. That is a step away from the past, where they had very firm ideas about it.

FIAF only tells you not to use household plastic bags, paper or other materials in the container. Out of 69 references, not a word about Manchester Polytechnic.

SMPTE allows a greater variation and very high temperatures, but a problem is that below 30% RH the film dries out and shrinkage occurs. The storage enclosures should preferably be tinned steel reels and cans or a variety of plastic types such as polystyrene, polyolefins and polyacrylate. That is a heavy mouthful that we will discuss later.

EBU allows 50% RH and 15°C. and the film can is not debated only to place the film within a plastic bag in the box.

2.5.1 Access to storage Material in the light of History.
Ever since film was introduced a hundred years ago it was manufactured and put into a can. The raw stock was supplied in metal tins in 120 meter and later 300 meter sizes. Very easy and reusable after development. When "One Reelers" 1000', which is 305 meters were replaced by 2000', which is 610 meters they were taken on for storage and happened to follow the film through it's entire life. Nice, easy and very cheap.

That is why most film storage enclosures are metal tins and that counts for most of the world. The Americans have used steel reels for 35mm and kept them in heavy duty storage containers. Funny way to save money.

Now, everybody faces the problem of raw stock on 4000'.!!

TV stations have acted somewhat differently. Apart from the fact that all film material must be readily available (no FICA!) Especially in the Nordic countries we find the use of cheap cartons for film storage.

2.6 Research in Manchester
Manchester Polytechnic has researched into the degradation and stabilisation of film and has been awarded a major research grant by the National Film Archive in London and Eastman Kodak, USA to investigate the problems of disintegrating archival film material and have come up with a solution.

This has been going on for 5 years. Many papers have been presented and reports written.

2.6.1 Conclusions from the Work to Date.
The nature of the storage enclosure is important. Tin coated or polymer coated metal cans are unsuitable. Iron is not to be used under any circumstances since it catalyses the oxidation processes in the polymer. Aluminium is better but again a weak catalyst. Paper/cardboard containers are unstable themselves and not to be used. They will generate acids on storage that can be damaging over a period of time.

Plastic containers are the most suitable from an economic point-of-view, but again the nature of the container is important. Normal polyolefin cans in use are unsuitable and need to be changed regularly.

The problem is that manufacturers of these cans don't believe they need stabilisation because they remain in the dark at room temperature. Here the stabilisation technologies used normally only involve a weak base stabilisation of a hindered phenolic antioxidant and a phosphite at a sufficient level to protect the polymer from degradation during its high temperature processing and fabrication into a can.

Such a formulation is insufficient to deal with the adverse effects of very long term storage of materials that themselves will degrade and then attack the can.

2.6.2 Humidity and Temperature

Environmental conditions for archival storage should be 15°C for triacetate film material in black & white. They should be stored at 30-35%RH and at temperatures below 15°C if possible. Humidity is the overriding factor with this material. Colour material must be stored at even lower temperatures due to the instability of the dyes, almost certainly below 5°C.

Polyester film base material is presumably archival from the results of our investigations but again polyester is very sensitive to moisture. Below 40%RH hydrolysis is negligble. Video tape and audio tape materials are unstable and not archival. This is due to the instability of the polyester-urethane coatings used. The binder has high thermal instability and here stabilisation technologies need to be developed by the manufacturers.

2.6.3 Storage enclosures

Apart from glass (basic type) the polyolefins are the most suitable and economic materials for archival containers since they are simplistic in their structure with only carbon and hydrogen atoms in their molecular make up.

Thus, high density polyethylene and high crystallinity polypropylene are the best materials since they act themselves as effective barriers to oxygen diffusion.

They need to be stabilised not just for processing but for long-term storage and to combat any changes arising from the film deteriorating inside. In this regard formulations utilising synergistic combinations of additives offering good thermal protection are needed.

The flame retardant is also important and should not be a halogen type since they may effect the film. In the long run metal based oxides or phosphates are preferable.

3. Conclusion

It is indeed very difficult to conclude, since so much is still being researched. There is little doubt that plastic film containers will be accepted more and more. It is said that in the United States the Library of Congress and others have started to change to plastic with some of the improved specifications.

I myself as a manufacturer of plastic film containers under the trade name of DANCAN have taken advantage of the research and added the necessary additives to stabilize DANCAN.

You still can't get something for nothing, and it adds to the price, but in view of the superior quality and expected lifespan, it is surely worth it.

On my initiative Manchester Polytechnic is doing extensive tests on several plastic film containers as a research project and I expect results by Easter 1992.

Bibliographic References

(1) Allen, N.S., Edge, Michelle., Jewitt, T.S., Horie, C.V.,: Manchester Polytechnics Centre for Archival Polymeric Materials : Compiled Report from 1989.

(2) SMPTE Journal., Volume 100, Number 7, July 1991.
Proposed SMPTE Recommended Practice RP 131.

(3) American National Standard. ANSI PH1.43-1985

(4) European Broadcasting Union, Tech 3202-E, Storage of magnetic tapes and cinefilms. August 1974.

(5) Schou, Henning., Preservation of Moving Images and Sound. Fiaf Preservation Commission 1989

Paper Conservation: Some Polymeric Aspects

D. J. Priest

DEPARTMENT OF PAPER SCIENCE, UMIST, P.O. BOX 88, MANCHESTER M60 1QD, UK

1. INTRODUCTION

There is no doubt that the cultural heritage of the world is to a great extent associated with paper. Although there are obvious areas of particular import- ance, like books, manuscripts, and works of art on paper there are a surprising number of other uses for paper, which are of historic value. To mention some, there are maps, terrestrial and celestial globes, eng- ineering drawings, postage stamps, wallpaper, and posters - not to mention such ephemera as theatre tickets and programmes, rail and bus tickets, and grocery store till rolls. All this represents a major problem for those concerned with conservation and pres- ervation, not least because each kind of historic object tends to be made from its own type of paper or paperboard. It must be recognised at the outset that a key characteristic of paper is that, although it has a common basic structure, it takes on a great variety of forms - consider for example, the apparent difference between a lens tissue, a 3 mm thick mounting board and a newspaper. Furthermore, there are less obvious diff- erences between grades of paper and board (the industry term for 'card-board') such as, amongst others, the nature of the fibre used, the amount and type of min- eral filler, and the degree and kind of sizing used.

As we shall see, some of these variations in cons- titution play a vital role in how paper changes with the course of time.

Paper and related products, perhaps because they have been around for so long and are commonplace and diverse, are rarely considered in the same context as newer polymeric materials such as nylon or polyester. However, despite key differences, to be enlarged upon below, paper is basically a polymeric material - it could, in fact, be described in modern terminology as a highly anisotropic polymeric composite. Also, many of the additional materials incorporated in various types of paper to modify its purpose are polymeric in nature.

In what follows, the nature of paper will be outlined from a polymeric standpoint. The reasons for
problems with the slow deterioration of paper will be
discussed, and how these are related to adjuncts used
in papermaking. Conservation treatments using polymeric material will be illustrated by one example concerned with consolidating and protecting soft pencil
marks. Finally, an outline is presented of current
developments in the area of mass treatments for books,
one of which involves polymerisation as a key feature.

2.   STRUCTURE AND PROPERTIES OF PAPER

2.1 Cellulose - the basic polymer

Cellulose is the polymer which is the basic structural
unit in paper. Its chemical structure is shown in
Figure 1, in a form which illustrates the 'chair'
conformation adopted by the anhydroglucose rings.

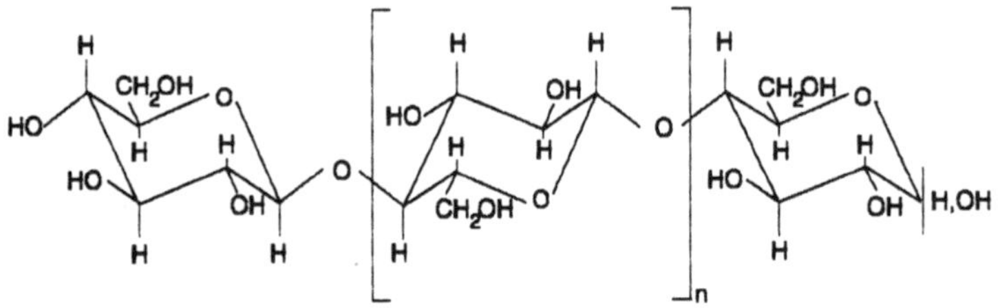

Figure 1. Chemical structure of cellulose, illustrating the
"chair" conformation adopted by the anhydroglucose rings.

In conjunction with the geometry of the 1-4 β links
between the rings, this allows cellulose to form
extended molecules which are very near to being linear,
and which can certainly form highly ordered crystalline
regions. The presence of three hydroxyl groups per
glucose residue, together with the oxygen atoms, make
possible both intra and inter molecular hydrogen bonding of various types. In addition to the crystalline
cellulose encountered in nature (known as cellulose I),
various other forms of crystalline cellulose are known,
all involving extensive hydrogen bonding. Although
weaker than a covalent bond, H-bonds are much stronger
than the intermolecular forces operating in less polar
polymers such as polythene or polypropene. This factor, coupled with the high degree of crystalline order
give solid cellulosic materials very different properties from 'conventional' synthetic thermoplastic
polymers. In fact, cellulose based materials, such as
paper, are not generally discussed in texts and treat-

ises on polymers and polymerisation.

The distinctive features of cellulose are, in the main, as follows:

a) There are no sharp distinctive glass transition or melting temperatures. As the temperature increases to around 170 - 180°C, cellulosic materials tend to chemically degrade, turning brown and charring, without any clear cut change in physical properties. Values are sometimes quoted for a glass transition temperature, but the transitions are certainly much less clear than in conventional polymers.

b) The capacity for forming large numbers of hydrogen bonds makes cellulose potentially very hydrophilic. Cellulose based materials, which are normally partly amorphous - or at any rate contain regions less well ordered than the crystalline ones - readily absorb moisture from the atmosphere. Under ordinary conditions, for example, paper will contain some 7 - 9% w/w of water. The mechanical properties of the materials depend markedly on the moisture content, including those indistinct transitions mentioned above. Indeed, it might readily be supposed that cellulose would be soluble in water. In fact, the high degree of extended H-bonded order in the crystalline regions ensures that insufficient water is able to penetrate these regions to form a solution and, again unlike conventional polymers, cellulose is insoluble in common solvents, both polar and nonpolar. Cellulose will dissolve in some unusual solvents, such as hydrated N-methylmorpholine-N-oxide, and in certain metal complexes [1], which does enable solution properties to be studied, but not in as direct a way as for other polymers.

These distinctive features of cellulose have, as we shall see, an important bearing on the long term stability and conservation of paper.

2.2 Supra molecular structure

The common feature in paper and board is the presence of a layered 'mat' of overlapping fibrous elements, bonded to a certain extent where they cross. This is well seen in Figure 2, a scanning electron micrograph showing surface structure. The fibres are nearly always of natural vegetable origin, and, for most grades of paper commonly encountered, are derived from wood. Being a natural product, it is hardly surprising that the basic polymeric cellulose chains are arranged in a complex manner in a typical wood fibre or tracheid. It is thought that a few individual polymer chains are assembled in fine 'microfibrils', which are

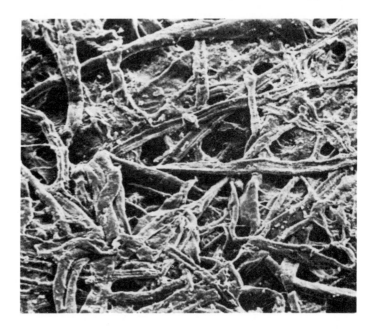

0.1 mm

Figure 2. Scanning electron micrograph of the surface of paper, showing the overlapping flattened fibres, in this instance derived from wood.

then arranged, in a softwood tracheid, in the manner illustrated in Figure 3. There has been much discussion about the exact nature of microfibrils and the supra molecular structure, and the reader is referred elsewhere for more detail [2].

2.3 Other Constituents

In most papers, substances other than cellulose are present in the fibrous structure or associated closely with it; many of which are polymeric to a greater or lesser extent. In papers containing 'mechanical' pulp - that is to say, pulp in which the tracheids in the wood are separated from one another principally by an internal mechanical shearing action of one sort or another - the major secondary material will be lignin. Lignin has a very complex three dimensional polymeric structure, based on alcohols with a phenyl propane skeleton; in the tree it serves to cement the cellulosic tracheids together and thus contributes greatly to the mechanical strength of the wood.

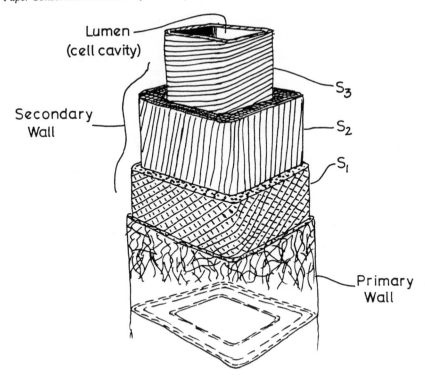

Figure 3. Sketch of the fine structure of a softwood tracheid, showing, schematically, the orientation of cellulose chain molecules in the various layers.

Paper containing lignin is used for such purposes as newspapers and paperback books; it readily yellows on exposure to light and it generally has a bad reputation amongst conservators, a matter which is discussed further below. Higher grade products contain paper in which the fibres have been separated out by a chemical treatment which also removes most of the lignin.

There are other important constituents which may be an integral part of the fibre, or may be added deliberately. In the former category, there are normally present short chain polysaccharides known as 'hemicelluloses'; not a good name because the chains contain a mixture of monomer units, many of which are non-glucosidic. These low molecular weight polysaccharides are thought to assist in obtaining good bonding between fibres. Many different kinds of materials can be added deliberately, and a full discussion is beyond the scope of this article. Some may be added to modify the properties of the paper, such as sizes, and mineral

fillers and coatings. Other additives are for controlling aspects of the production process and are often
polyelectrolytes, an example being a 'retention aid' to
increase the proportion of added mineral filler
retained in the sheet. Some aspects of permanence and
the functional type of additive will be mentioned
below.

2.4 Mechanical Properties

A great concern to those involved in the preservation
and conservation of paper is loss of strength on
ageing. On storage for long periods, paper can become
quite brittle, so that, for example, books can no
longer be used without pages breaking when they are
handled.

Before looking at the reasons for this, it is
important to have some knowledge of the nature of mechanical strength in paper. A typical load elongation
curve for paper is shown in Figure 4. A short elastic
region is followed by yielding and failure. Elongation
at break is much less than for a rubbery polymer, and
more like that for a rigid polymer, despite the relative flexibility of a sheet of paper.

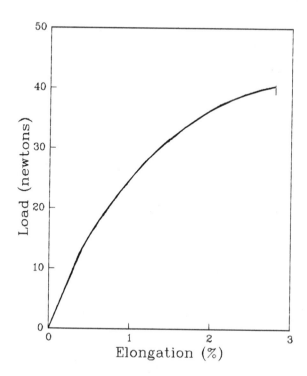

Figure 4. A typical load–elongation curve for paper.

The load at the break point is used to derive a tensile strength for paper, usually expressed as a load per unit width, and which may be corrected for variations in grammage, or mass per unit area, of different grades of paper. The tensile strength can be thought of as arising partly from the strength of individual fibres in the sheet, and partly from the strength of fibre-fibre bonds (refer to Figure 2). The strength contributed by bonding between fibres can be affected by many factors including the area in contact, the proportion of that area which is bonded, the strength of the bond, the perimeter of the fibre cross section, and the average fibre length. A means of measuring fibre strength alone is afforded, to a good approximation, by carrying out a tensile test in a device which clamps a strip of paper leaving no unclamped paper between the jaws. This is known as a 'zero span tensile test', in comparison with a finite span test in which the strip of paper between the jaws normally measures at least 100 mm.

The various parameters have been combined in equations, and a generally respected version is that due to Page [3], which can be simplified to

$$\frac{1}{T} = \frac{k_1}{Z} + \frac{k_2}{b}$$

when T is the finite span tensile strength
Z is the zero span tensile strength
b is the bond strength
k_1 and k_2 are quantities containing the other variables.

When a paper ages and becomes brittle, the shape of the finite-span load elongation curve changes, as shown schematically in Figure 5. In general, there is proportionately less change in tensile strength (load at break) than in elongation. It can also be seen that there is a slight increase in Young's modulus, as shown by the increase in slope of the early elastic region of the curve [4]. A useful parameter for characterising the 'toughness' of a paper is the tensile energy absorption (TEA), which is the total energy absorbed in breaking the sample, and is given by the area under the load elongation curve. As Figure 5 shows, and as might be expected, 'brittle' paper has a much reduced TEA.

In contrast to the finite span tensile strength, the zero span figure decreases markedly as the paper embrittles. This is illustrated in Figure 6, in which the two values are compared for the same paper. This result, and others used as illustration in this article, was obtained in the author's laboratory. Extensive use has been made of the pilot scale papermaking machine available in the laboratory, which enables paper to be made with known properties, and containing only those constituents essential for the experiment in hand, thus avoiding unnecessary

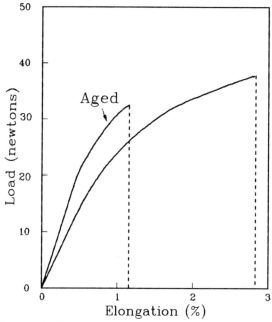

Figure 5. Diagram showing how ageing usually affects the form of the load—elongation curve.

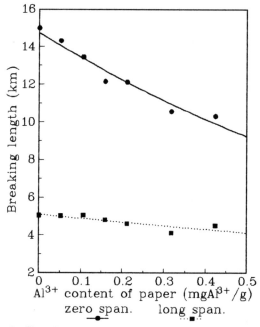

Figure 6. The decrease of zero span and finite span tensile strength as paper degrades due to acid hydrolysis. Samples were all aged for the same time; the variation in degradation being produced by treating the papers with varying amounts of aluminium sulphate.

complications. The paper used for the tests illustrated in Figure 6 was treated in the laboratory with
solutions of aluminium sulphate in order to produce
different degrees of degradation after a standard
period of accelerated ageing - this aspect is discussed
more fully below (see paragraph 3.2).

The results indicate that on degradation, fibre
strength is affected to a much greater extent than the
strength of the bonds between fibres. Referring again
to the simplified Page equation, in a given sample of
paper it can be assumed that geometric factors such as
fibre length and perimeter etc will remain constant as
ageing proceeds. If it is assumed that bond strength
also remains constant, a graph of 1/Z against 1/T
should give a straight line, and this is illustrated in
Figure 7.[5]   The one point off the line represents a
very highly degraded sample where other effects may be
playing a part.

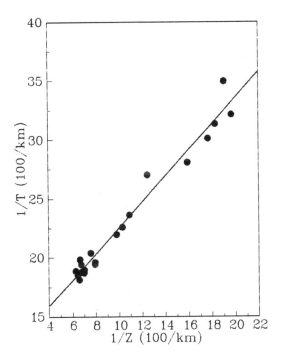

Figure 7. A graph of the reciprocal of zero span tensile
strength against the reciprocal of finite span tensile
strength, indicating that embrittlement is predominately due
to loss of fibre strength rather than inter fibre bond
strength (see text, para. 2.4).

We now need to consider the reasons for this loss
of mechanical strength, but before doing so it is necessary to look further at the additional constituents of
paper.

3.   ADDED CONSTITUENTS IN PAPER

As mentioned earlier, there are many possible non-
cellulosic components and added materials. One type is
of particular importance in considering degradation,
and that is sizing. A comprehensive treatment of other
additives is outside the scope of this article, but
some comments will be made on a few other important
non-cellulosic materials.

3.1 Sizing

In modern papermaking, sizing means primarily providing
the paper with resistance to wetting and penetration of
aqueous fluids such as writing inks, adhesives, or
aqueous based coating preparations. The term is also
used for treating paper to improve, for example, the
ability of surface fibres to resist being 'picked' out
during printing or other conversion processes. The
provision of water resistance is called 'internal sizing' whilst treatment of surfaces is termed 'surface
sizing'. In fact, the two processes are normally alto-
gether distinct, as we shall see.

Paper made by hand, before the advent of papermaking machines in the early 19th century, was often
treated by immersing the whole sheet in a bath of gelatine solution, and this process, which improved both
the surface characteristics and sealed surface pores,
was also known as sizing. It carries over into machine
made paper in the form of 'tub sizing' in which the web
of paper passes through a solution, often gelatine, the
excess being removed by squeeze rolls before the paper
is re-dried. This process is very seldom used nowadays,
except for very specialised grades.

The distinction between the various forms of sizing needs to be clearly understood, especially since
conservators are often involved with early types of
gelatine sized (i.e. tub-sized) hand made papers, which
are quite unlike a modern   'internally sized' grade of
paper.

3.2 Rosin/alum sizing and paper degradation

It is internal sizing which is of particular rel-
evance to paper degradation. The commonest method used
was introduced early in the 19th century; it has been
much developed and refined since then but remains basically the same. Very briefly, it involves adding a

'rosin' plus aluminium sulphate (known to papermakers as 'alum') to the dilute suspension of fibres in water from which the continuous web of paper will be formed. The individual fibres in the dried paper are then found to be hydrophobic, since they carry on their surfaces a very thin coating of the rosin.

The rosin is derived from trees, one of its major constituents is abietic acid, Figure 8. This carboxylic acid is capable of forming soaps with alkalies, which renders it miscible with water, producing a negatively charged moiety (In practice, the rosin can be used in various forms, but this is not relevant to the present discussion).

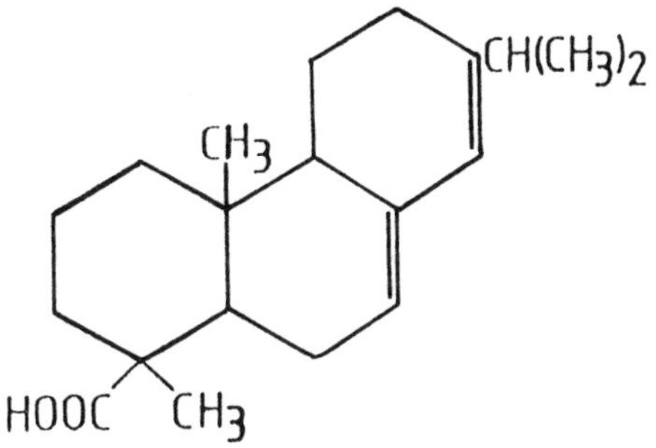

Figure 8. Abietic acid

Since the surfaces of the dispersed cellulosic fibres tend also to be negatively charged, some way must be found for overcoming the repulsive forces between the fibre and the size. The answer is the use of the aluminium ion, which, normally having three positive charges and being relatively small in size, provides, in some way not yet fully understood in detail, a link which binds the size particles onto the fibre surface. Since size and rosin are both inexpensive, this is an efficient and cost effective way of producing the necessary resistance to wetting.

However, since aluminium sulphate is a salt of a strong acid and a weak base, when it is dissolved in water hydrolysis takes place and the solution becomes acidic. The following reaction shows how one of the possible products of hydrolysis is formed by interaction between the hexa-hydrated aluminium ion and water:

$$Al(H_2O)_6^{3+} + H_2O \rightleftharpoons Al(H_2O)_5OH^{2+} + H_3O^+$$

Various other hydrated hydroxy-aluminium ions are possible, including some which are polymers containing

a number of aluminium atoms. The nature of the ions in
solution depends on concentration and pH, but the net
effect of simply dissolving aluminium sulphate in water
at around the concentrations used in papermaking is to
give an acid solution with a pH of 4 - 4.5. We have
studied this effect in a simple system using our pilot
papermaking machine. It appears that once the alum is
dissolved, the aluminium ions and acidic oxonium ions
(H_3O^+) act independently. As the concentration of alum
increased, the paper being made became steadily more
acidic (ie an aqueous extract of the paper showed a
steadily decreasing pH), but the uptake of aluminium
remained at the same low level [5].

Unfortunately, the presence of oxonium ions in
paper leads to acid hydrolysis of the basic cellulose
polymer molecules, resulting in chain scission at the
1-4ß links between the anhydroglucose molecular units
(cf Figure 1); this is a well documented reaction [6]. A
rapid fall in degree of polymerisation results, which
can be followed by viscometric measurements in a suitable complexing solvent such as cadoxen,
tris(ethylenediamine) cadmium (II) hydroxide. Figure 9
illustrates this for the same papers used to give the
tensile strength values shown above in Figure 6. It
will be seen that the form of the decay in strength is
different from that of degree of polymerisation, which
is hardly surprising in view of the complex supra molecular structure in the fibre.

It is highly probable that this acidic deterioration, arising from alum-rosin sizing, is the most
serious source of the huge problem with 'brittle books'
facing repositories world-wide, especially by the major
national copyright libraries.

These are certainly other possible sources of
acidity in paper, such as acid gases in the atmosphere
and residues in the cellulose structure, refer elsewhere for a full review [7]. However, it needs to be said
that not all these sources of acidity are likely to
cause deterioration at the same rate, and whilst all
sources must be taken into account where the longest
possible lifetime for a paper is required, alum rosin
sizing remains the single greatest worry for existing
paper.

3.3 Cross-linking reactions

Use of acidic aluminium solutions in papermaking can
also lead to minor effects thought to be associated
with some kind of polymer cross linking. This shows up
most strikingly in tensile tests performed on wet
paper, because in the absence of cross links, water can
penetrate the inter fibre bonds and the paper loses
virtually all its strength. Figure 10 and Figure 11
show the effect of thermal treatment in the presence of

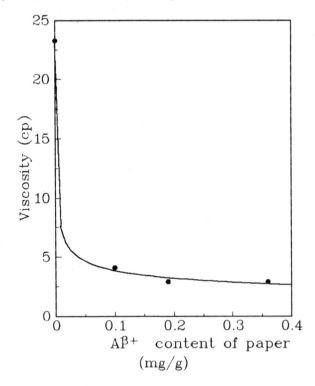

Figure 9. The effect of degradation on the viscosity of cellulose solutions.

aluminium ions [8]. Zero span tests show the expected degradative loss of fibre strength due to depolymerisation, but, in direct contrast, long span tests show an <u>increase</u> in strength attributable to water resistant cross linking in fibre-fibre bond regions. In 'dry' paper (ie paper under ambient conditions) absolute values of strength are naturally much higher, and the effect is more difficult to see. However, it may contribute to the observed slight increase in modulus often observed in aged paper (see Figure 4 and Ref 4).

3.4 Non-acidic methods of sizing

There are a number of economic advantages to be gained by making paper under neutral or alkaline conditions. For example, calcium carbonate filler (which reacts with acids) can be used in somewhat greater quantities and this reduces costs. These economic incentives have resulted in the introduction of internal sizes which are designed to react chemically with accessible hydroxyl groups on the surface of the fibres. One such size is based on an alkyl ketene dimer (AKD); Figure 12 shows how AKD is thought to react with cellulose.

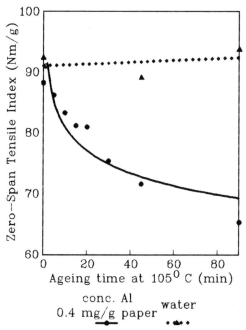

Figure 10. The change in wet zero span tensile strength on thermal treatment of paper, showing the degradation produced by impregnation with aluminium sulphate solution.

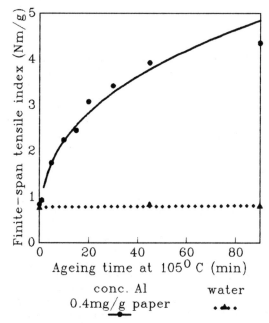

Figure 11. The change in wet finite span tensile strength on thermal treatment of paper; compare Figure 10. Impregnation with aluminium sulphate solution has caused an <u>increase</u> in wet strength due to enhanced inter fibre bond strength.

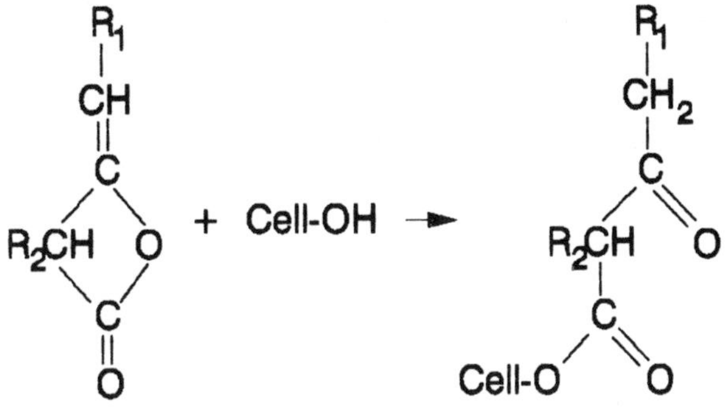

Figure 12. The postulated reaction of alkyl ketene dimer (AKD) size with cellulosic hydroxyl groups. R_1 and R_2 are normally short alkyl chains.

Whilst not highly polymeric, the hydrophobicity derives from $C_{16}-C_{18}$ alkyl chains. The reader is referred elsewhere for more details [9].

The rate of accelerated ageing of rosin/alum and reactive (AKD) sized papers is compared in Figure 13. Once again, the papers were made on our pilot scale machine and were identical apart from the type of sizing. The rate is shown in terms of folding endurance, plotted as the logarithm of the number of double folds to failure; this test is rather imprecise but very sensitive to degradative loss of strength. Clearly, the AKD sized paper is losing strength at a very much slower rate than the acid sized paper. In fact, the pulp suspension being supplied to the paper machine was still slightly acidic, because in keeping with the experimental design, the natural slight acidity of the wood pulp slurry was not neutralised - as it would have been if an alkaline filler like calcium carbonate had been used.

Although neutral/alkaline sizing was introduced entirely for economic reasons, it has had a profound effect on future prospects for the longevity of paper intended for high volume use in books and documents. (It has long been possible to obtain highly priced long-life papers for special archival applications). One consequence is the formulation of national and international standards for paper for printed library materials and similar applications [10]; an ISO standard is due to be published soon. For a fuller account of neutral/alkaline papermaking and its impact on permanence, reference should be made to the writer's chapter in a recent American Chemical Society publication [11].

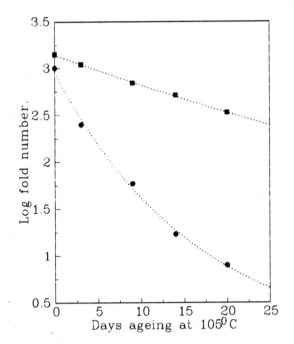

Figure 13. Comparison of the rate of ageing for the same
type of paper, sized with either rosin/alum or AKD.

3.5 Lignified Papers

Although not strictly an additive, it is convenient at
this point to discuss some aspects of the longevity of
papers containing the complex polymer lignin (cf para
2.3). Conservators are most likely to encounter these
papers in the form of newspapers, paperback books or
posters; such papers generally have a reputation for
very rapid degradation. They certainly rapidly discolour and turn yellow/brown, especially on exposure to
light. The discoloration can readily cause staining on
adjacent sheets of paper. Since these types of paper
are generally weak to start with, even a small amount
of degradation of physical properties is detrimental.
Lignin can, it is claimed, produce acidic residues on
degradation which cause chain cleavage; however, the
extent to which this occurs is not clear and the
discoloration is not always matched by a marked
decrease in strength.

Pulps in which the lignin is retained are known as
'high yield' pulps, in that a very high proportion of
the wood (95% or more) is converted into paper, and
this is clearly advantageous to manufacturers. Modern

methods of producing high yield pulps enable stronger papers to be made, which are often bleached to a good degree of whiteness. One such pulping process is called 'chemithermomechanical' (CTMP) in which the wood is broken down mechanically at relatively high temperatures and pressures in the presence of low concentrations of pulping chemicals. Papers from CTMP pulp can compete with office and stationery grades which currently would contain little or no lignin. We need to know how these new types of paper degrade, and, once again, we have used our pilot paper machine to make suitable papers for test. On the whole, it seems that papers containing CTMP will retain strength satisfactorily on ageing provided they are sized using a neutral/alkaline system and contain calcium carbonate as a filler; some results are known in Figure 14, strength being measured in terms of folding endurance. However, the presence of substantial properties of lignin does mean that the papers will rapidly yellow on exposure to light.

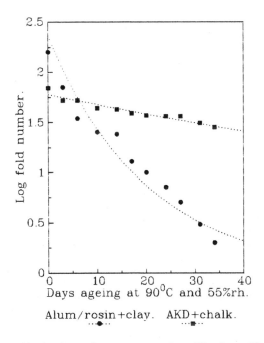

Figure 14. Ageing of papers made with chemithermomechanical (CTMP) pulps, showing the dominating effect of the type of sizing system and filler used.

3.6 Effect of pigment coatings

Whilst the presence of surface layers of pigment (clay or calcium carbonate) might not in themselves at first sight be expected to affect the rate of degradation of the underlying paper, coatings contain other materials, and there is always some degree of penetration of the liquid coating mix into the paper. Our research in this area, reported in detail elsewhere [12], has shown that coatings can affect the rate of ageing (Figure 15). One possibility is associated with the use of polymeric adhesives in the coating preparation. Commonly used adhesives are emulsions based on polymers of vinyl acetate or copolymers of styrene-butadiene; when these materials penetrate the paper they appear to modify in some way the process of degradation.

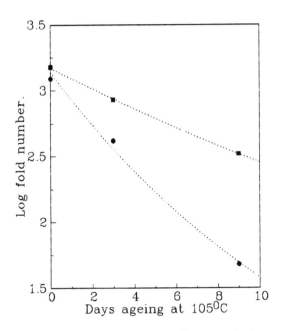

Uncoated paper.   Clay coated.
       ···●···   ··■···

Figure 15. An example of the effect coating can have on the rate of ageing.

4   POLYMERS IN CONSERVATION TREATMENTS - AN
     EXAMPLE

A very wide range of polymeric materials are used in
the conservation of paper and paper board artefacts,
many of them cellulose derivatives. A full treatment
is outside the scope of this presentation, and for
further information the reader is referred elsewhere [13].
As an example, an interesting application of a modified
starch will be described briefly; this has been the
subject of some recent research in the author's laboratory [14]. Starch is an abundant natural polysaccharide
with a chemical structure very like cellulose. The
vital difference is that the monomeric anhydroglucose
rings are connected by α 1,4 links instead of the ß 1,4
arrangement in cellulose. Also, as well as a linear
(unbranched) 'amylose' chain, starch contains a proportion of a high DP branched amylopectin - still based
entirely on the glucose unit, but branched via α 1,6
links. The α 1,4 link in the main chains has a major
effect on the geometry of the polymer molecules and how
they fit together and hydrogen bond to one another.
Consequently, when starch grains are heated in water to
temperatures around 70 - 80°C they swell and burst,
forming a translucent highly viscous dispersion or
'paste'. On cooling to room temperature, this dispersion sets to a gel, which does not re-disperse when re-
heated. The gel contains 'bundles' of well ordered
amylose molecules, which act as stable cross links in
the gel structure. Dispersions of starch which are
free flowing at room temperature can be produced by
various means of modification often involving a controlled extent of depolymerisation. These modified
starches are widely used in the paper industry for,
amongst other things, surface sizing and adhesives for
coatings.

A modified starch has now been found to have a
useful application as a fixative for manuscripts written using very soft pencil. Particularly valuable
examples, which gave rise to the research, are the
Shelley notebooks in the collection of the Bodleian
Library. The key advantage of the modified starch is
that dilute solutions can be readily sprayed on to give
just sufficient a surface layer to fix the graphite
fragments. Experiments involved sealing dummy pencil
marks, and using an abrasion tester to partially transfer the marks to a contacting paper pad - the amount
transferred being determined by optical reflection, or
'brightness'. Some results are shown in Figure 16. A
small addition is seen to greatly reduce the amount
rubbed off. Enzyme treatment was found to be a convenient, gentle, safe and controllable method of modifying raw starch to give suitably free flowing
solutions. Tests on commercially modified starches,
used for size press treatment on our pilot paper
machine, showed that they had no effect on the rate of

accelerated ageing.

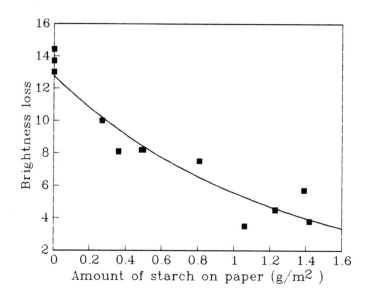

Figure 16. Graph showing how surface treatment with enzyme modified starch reduces the tendency for soft pencil marks to rub off onto neighbouring sheets of paper. (Less rub–off equates to a reduced brightness loss).

5   MASS TREATMENTS FOR BOOKS AND DOCUMENTS

The sheer scale of the brittle paper problem affecting major copyright libraries has led in recent years to a number of proposals for treating books on a large scale. In the British Library, for example, some 7-8 million volumes [15] are in a poor condition - a number which it is completely impracticable to treat by conventional labour intensive methods of deacidification. In an ideal mass treatment process, a batch of some 100 books or so would be placed in a container, pass through an (unspecified) treatment plant, and be returned to the shelves to all intents and purposes completely unaffected, except that they would have become virtually everlasting - and possibly stronger! No such process yet exists in all these aspects, but it is valid to separate the concepts of deacidification, ie arresting the progress of degra-dation, and strengthening, ie making a brittle book usable again - something which deacidification on its own cannot be expect to do.

Various schemes are at various stages of develop-ment. Four such schemes have been selected for comment here, they are probably the ones most likely to be used

more widely in practice. One of them makes use of
polymerisation technology, and the writer has been
associated with it - it is dealt with first although it
has not reached the same stage of development as the
others.

5.1 Internal Polymerisation - the British Library Process

Research for this process was carried out at the
University of Surrey, sponsored by the British
Library[16]. It has been found possible to infiltrate a
mixture of acrylic monomers into the interior structure
of the pages of a book, and to initiate polymerisation
by exposure to γ-irradiation. Surprisingly, the pages
do not stick together, the micro-roughness of paper
being presumably sufficient to prevent this.
Substantial increases in strength can be obtained from
this polymeric reinforcement, and this is the major
effect observed. A degree of basicity, and hence ant-
acid actions, can be introduced by using an appropriate
co-monomer. However, the best effect is produced by
combining a deacidification treatment with the internal
polymerisation. This is shown in Figure 17, results
from the writer's laboratory, again using paper
specially made on the pilot paper machine.

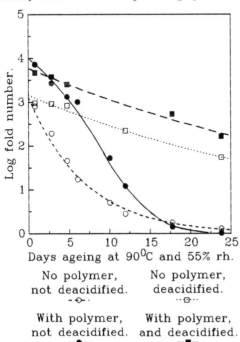

Days ageing at 90°C and 55% rh.

No polymer,   No polymer,
not deacidified.   deacidified.
- ◇ -   ·· ▫ ··

With polymer,   With polymer,
not deacidified.   and deacidified.
—●—   —■—

Figure 17. The results of accelerated ageing tests
on papers treated by the British Library polymer
strengthening process.

The sharp increase in strength can be seen, and the
effect of deacidification. The British Library now
hope to be able to take the project to the pilot plant
stage.

5.2 The Union Carbide 'Wei To' non-aqueous system.

In this process, books are immersed in an inert solvent
(fluorohydrocarbons have been used) containing an
organic base which has first to be dissolved in a polar
solvent such as methanol [17]. An example of such a base
is methoxy magnesium methyl carbonate, which hydrolyses
to yield magnesium hydroxide, which in turn acts as the
neutralising agent and, in excess, provides an alkaline
reserve. The water for the initial hydrolysis step
needs to come from residual moisture in partially dried
books.

$$CH_3-O-Mg-CO_2-OCH_3 + 2H_2O \rightarrow Mg(OH)_2 + CH_3OH + CO_2$$
$$Mg(OH)_2 + H_2SO_4 \rightarrow MgSO_4 + 2H_2O$$

This process has been in use in the National
Archives of Canada on a modest scale for some 10 years.
As it is operated there, some pre-selection of books is
needed because the solvent system can interact with
some dyestuffs and colorants. Consequently the process
does not meet the 'ideal' conditions described above,
although it has proved a successful means of increasing
the rate of treatment beyond that practicable by hand.
This process is entirely for deacidification; there is
no strength enhancement.

5.3 The 'DEZ' process.

This process has the distinction of being a vapour
phase treatment, which is attractive because it completely eliminates the need for books to be immersed in
a liquid of any sort. The active compound chosen is
diethyl zinc, hence the DEZ acronym. This is a very
highly active substance, reacting readily, and
exothermally, with water, oxygen, and acids:-

a. $(C_2H_5)_2Zn + H_2O \rightarrow ZnO + 2C_2H_6$
b. $(C_2H_5)_2Zn + 7O_2 \rightarrow ZnO + 4CO_2 + 5H_2O$
c. $(C_2H_5)_2Zn + H_2SO_4 \rightarrow ZnSO_4 + 2C_2H_6$

Consequently, books must be placed in a high vacuum chamber from which all oxygen is excluded. A slight
residual moisture content is needed to ensure that some
zinc oxide remains in the paper to act as a potential
alkaline reserve (equation a). Traces of oxygen are
dangerous, because reaction b is highly exothermic and
can be explosive.

The initial work on this process was carried out
by the US Library of Congress [18]; Akzo Chemie have now
built a pilot plant in which trial quantities of books

are being treated. Some doubts have been expressed about the introduction of the extraneous zinc ions into the paper, because for example, compounds such as zinc chloride which is hygroscopic might possibly be formed; these, and other factors, will no doubt be assessed during pilot plant trials. Once again, this is a deacidification, not a strengthening process.

5.4 The Lithco Process

This process, developed relatively recently by a subsidiary company of the FMC corporation, involves the use of a novel alkyl organic base which is directly soluble in a completely inert solvent; this should avoid any problems with labile dyestuffs and so on. The new base is derived from magnesium carbonate, having two adduct chains, each consisting of an outer short alkyl section and a short alkoxy section coordinated to the magnesium carbonate. On contact with residual water in the paper, the molecule is hydrolysed to magnesium hydroxide, to act as the neutralising agent, plus two alkyl/alkoxy chain molecules. These are thought to be able to enhance the mechanical properties of the paper somewhat, although the mechanism is not clear and the magnitude of the effect is likely to be less than that obtained by internal polymerisation.

A feature of this development is the attention being paid to the logistics of the whole process - from books on shelves to books being replaced on shelves - and marketing aspects in general. A further novel aspect is the use of dielectric heating for initial drying of the batches of books before introduction of the treatment liquid.

ACKNOWLEDGEMENTS

The author gratefully acknowledges the help of Daven Chamberlain, Gerry Davison, Liu Juntai, Eileen Rodgers and Judith Stanley in providing data and the scanning electron micrograph.

REFERENCES

1.   R.A. Young and R.M. Powell (Eds), 'Cellulose;
      Structure, Modification and Hydrolysis', John
      Wiley & Sons, New York, 1986.

2.   J.F. Kennedy, G.O. Phillips and P.A. Williams
      (Eds), 'Cellulose; Structural and Functional
      Aspects', Ellis Horwood Ltd., Chichester, 1989.

3.   D.H. Page, TAPPI, 1969, 52, 674.

4.   E.L. Graminski, TAPPI, 1970, 53, 406.

5.   D.J. Priest and M. Farrar, 'The Effects of
      Aluminium Salts on the Degradation of Paper',
      Proceedings of Symposium 88, Canadian Conservation
      Institute, 1988.

6.   L.T. Fan, M.M. Gharparay and Y-H. Lee, 'Cellulose
      Hydrolysis', Springer-Verlag, Berlin, 1987.

7.   M. Hey, The Paper Conservator (Journal of the
      Institute of Paper Conservation), 1979, 4, 66.

8.   D.C. Chamberlain, PhD Thesis, University of
      Manchester, 1991.

9.   J.C. Roberts (Ed), 'Paper Chemistry', Blackie,
      Glasgow and London, 1991.

10.   D.J. Priest, 'Permanence - The Standards',
      Proceedings of the Annual Conference of the
      National Preservation Office, The British Library,
      London, 1990.

11.   D.J. Priest, 'Permanence and Neutral/Alkaline
      Papermaking' in 'Historic Textile and Paper
      Materials, Conservation and Characterisation II',
      American Chemical Society, 1989.

12.   D.J. Priest, J. Liu and R.L. Prosser, 'Pigment
      Coated Paper - Experiments on Permanence',
      Proceedings of the International Conference on
      Book and Paper Conservation, Budapest, 1990.

13.   R.L. Feller and M. Wilt, 'Evaluation of Cellulose
      Ethers for Conservation', The Getty Conservation
      Institute, 1990.

14.   N. Bell and D.J. Priest, The Paper Conservator
      (Journal of the Institute of Paper Conservation),
      1991, 14, 53.

15.   N. Seeley and N. Barker, Chemistry in Britain,
      1979, 15, 305.

16. E. King, <u>Library Conservation News</u>, 1986, <u>No.12</u>, 1.

17. R.D. Smith, <u>American Libraries</u>, 1988, <u>19</u>, 992.

18. P.G. Sparks, <u>Restaurator</u>, 1987, <u>8</u>, 106.

Early Advances in the Use of Acrylic Resins for the Conservation of Antiquities

Maureen Robson

CONSULTANT CONSERVATOR, STUDIO, 29 PARK AVENUE, BIRMINGHAM, B18 5ND, UK

ABSTRACT

During the post war years acrylic resins were widely used in the conservation of antiquities. The two main categories of these synthetic resins are thermosetting and thermoplastic of which the latter may offer long term reversibility, an essential prerequisite to all conservation treatments. Consolidation processes provide additional strength to fragile porous archaeological materials such as bone and terracotta, where a change in colour is undesirable. Surface applications will reattach crazed lacquer, glaze and decorative pigment, to japanned lacquerware, ceramics and polychrome wooden sculpture or consolidate friable surfaces after desalination of ceramic vessels and stone statuary. Some of the thermoplastic resins, in emulsion form with a continuous phase, will provide a suitable water soluble paint system which on drying becomes an acrylic film reversible in an organic solvent. During the mid seventies came the development of a one part acrylic system which fast cures under ultraviolet light and fulfils all the criteria to effect full restoration and conservation of antiquities.

INTRODUCTION

During the post war years, with the advent of cheap supplies of oil, the acrylic resins, polybutyl methacrylate (PBMA) and polymethyl methacrylate (PMMA) resins were widely used in the restoration and conservation of antiquities. The two main categories of these synthetic resins are thermosetting and thermoplastic of which the latter category has been proved capable of long term reversibility which is now considered an essential prerequisite to all conservation treatments. The tendency for thermosetting resins to cross link renders them other than user-friendly when restoration is required! If the tendency is proved , these resins would not be selected today for use in conservation. However, during the post first war years, both categories of acrylic resins were put to a variety of uses for the conservation of a wide range of antiquities.

CONSERVATION

ACRYLIC POLYMERS: SOLID OR IN SOLUTION

Among the early uses of the PBMA resins, such as Lucite 44, was as a varnish for oil paintings. However, in the 1930's and 1940's these conservation practices were not generally documented; so it was over twenty years before any such studies were published. (Werner 1952) The fact that PBMA showed a tendency to cross link when exposed to light negated its reversibility in the long term, although short term reversibility was achieved by its solubility in hydrocarbon solvents. It also had advantages of flexibility, colour and clarity retention both in the short and long term and was widely used as a picture varnish up until the 1950's.

By this time, long term reversibility was now regarded as an important criterion which led to the search for a more suitable resin. During the 1950's saw the emergence of Paraloid B72 used initially as a lacquer for silverware and copper alloy (Biek 1952) and subsequently put to a variety of uses in the field of conservation including the consolidation of fossils (Rixon 1955) and earth (Shorer 1964). Surface applications with various solvents and matting agents will attach crazed lacquer glaze and decorative pigment to japanned lacquerware, ceramics and polychrome wooden sculpture. Consolidation processes provide additional strength to fragile porous archaeological materials such as bone and terracotta, where a change in colour is undesirable. During the late fifties to early sixties a new approach was introduced with the idea of complete immersion in a synthetic resin. For the first time, account had to be taken of the possible effect of the conservation process, on the external appearance of the porous material such a bronze age pottery urns or excavated bone. Any testing or sample removal must be carried out prior to treatment or authenticity would be lost. The complete pottery urn, its sherds/or bone were immersed in a suitable solution under vacuum of Bedacryl 122X (polybutyl methacrylate) and later Paraloid B72 (acrylic polymer) in xylene. (Robson 1988).

Bedacryl 122X (PBMA) has had a chequered career. During 1955-56 it was thought well of by the British Museum's conservation scientists but fell from favour twenty years later when its long-term reversibility was in doubt. Currently, Bedacryl 122X (PBMA) is again in general use and can be purchased in two forms: either as a solid which can be dissolved to form the resin, or in a concentrated solution. Both forms may be diluted with toluene to the required concentration and when applied dries to a matt finish. These resins were effective as means of consolidation since they were used to strengthen whole vessels on both excavation and in the laboratory workshop before polyvinyl alcohol emulsions were available. They were also used on individual fragments which could later be re-assembled and bonded with a suitable adhesive. However, the resin treatment did result in a certain amount of discolouration of the pot fabric. (Shorer, personal communication) A vacuum desiccator is used for this purpose and a glass vessel of small dimensions is used as the container, in order to economise in the use of the solvent. This vessel is placed inside the desiccator which is then evacuated. The object is allowed to remain under vacuum until all bubbling has ceased so that the consolidant is absorbed by the porous pot fabric. Air is then admitted and when the desiccator reaches atmospheric pressure, the urn is lifted out of the container and allowed to drain: it can then be dried without acquiring a gloss by leaving it in an atmosphere of the solvent (Plenderleith 1962) Durofix and HMG (nitro cellulose adhesives) form satisfactory and compatible adhesives for repair work. The two adhesives are easily reversible in the short term by application of acetone. Paraloid B72 dissolves more slowly in acetone than Durofix or HMG, so it is possible to dismantle joints while leaving the consolidant intact. More recently Paraloid B72 (PBMA) has also been packaged in a tube format for ease as an adhesive for both textiles (Beecher 1968) and wood (Angst 1979).

Clear paper varnish (Winsor and Newton) was another favoured acrylic polymer used in the early conservation pioneer practices within the British Museum during the 1940's. Although it has not been on the market for many years, it mixed well with levigated fine powder (Winsor and Newton) and furnished a clear colourless smooth glaze to porcelain either by brush or spray application. It required two days to harden. A camel hair brush was used to spread out the varnish across the bond area so that by the time it reached the join the varnish layer was perceptively thinner, thus eliminating the risk of a shadow being formed from any hard edge from applying the varnish by brush. (Shorer, Personal communication)

ACRYLIC MONOMERS

The use of polymethyl methacrylate resins for the encapsulating of natural history (and other specimens) was practised from the 1940's until the early 1960's. (Organ 1963)

During the 1950's Technovit 4004a (Germany) and its British counterpart Tensol (ICI, UK) were both widely used for gap filling perspex and thus suitable as an application for restoring glass. However, both contracted considerably and their shrinkage in the short term, after three months and after three weeks respectively, seriously limited their long term suitability in the field of restoration. (Shorer, personal communication)

Cyanacrylates too, have been suggested for glass but as they break down in the alkaline conditions on most glass surfaces, they have met with only limited success. For use in combination repairs with epoxy resins, they locate and temporary hold the glass join when applied at intervals along the break edge, before the epoxy resin cures. (Terwen, 1983)

DISPERSIONS

The use of acrylic dispersions was favoured during the 1950's; the retention of their optical properties in the long term bettered the polyvinyl acetate dispersion applied films although there was some variation from batch to batch. These dispersions have been widely used for heat-set adhesives, liquid adhesives and consolidants for paper. Texacryl 13002 and Revacryl 1A (Acrylic copolymer emulsions) have more recently been used to attach fragile woven fibres to a support fabric of suitable weight where the adhesive is applied to the support fabric in the wet state and allowed to dry. The adhesive is then reactivated by application of gentle heat with a spatula iron. (Robson, 1986)

Some of the thermoplastic acrylic resins in emulsion form with a continuous phase will provide a suitable water soluble paint system which on drying becomes an acrylic film reversible in an organic solvent eg. Rowney Cryla Flow acrylic paints. These paint systems are light stable and widely used in the retouching of restored lacunae in a wide variety of antiquities ie ceramic, glass, stone, wallplaster etc.

During the mid seventies came the development of the one-part acrylic system which fast cures under ultraviolet light and fulfils all the criteria to effect full restoration and conservation of antiquities.

1. These products are one part systems that contain no solvents.

2. They are clear colourless liquid photopolymers which are suitable for application by syringing and will fast cure when exposed to ultraviolet light.

3. They may be applied direct to the prepared lacunae in association with the fractured enamel.

4. A short precure with ultraviolet light of sufficient duration to set the bond allows it to be readjusted, if necessary, without disturbing the alignment.

A comparative study of a range of glass adhesives was undertaken from 1983 - 1986 to establish a range of adhesives with the following criteria: retention of optical properties, non yellowing on ageing, reversibility both in the short and long term and stability under exposure to ultraviolet light and moisture.

Of the ultraviolet curing (photosensitive) acrylics tested, Norland optical adhesives (urethane related prepolymers) were found to satisfy such criteria. (Unpublished research, Robson, 1986) These UV curing acrylic adhesives with a thin consistency, can be syringed directly into cracks in situ, can be realigned after a short precure of ultra violet light of sufficient duration, to set the bond, and retain absolute colour and clarity in both the short and long term. The adhesion of NOA61 and NOA63 is good initially but will become greater with ageing over a period of about two weeks. Acetone is used to remove the uncured resin, whilst after curing any rejects can be separated in methylene chloride. The bond area must be soaked in the solvent and normally will separate over night. The time required to break the bond depends upon the extent of cure and area of the bond.

At present a suitable dispersant is sought to enable these acrylic adhesives to pass easily though a compressed air spray attachment. Should recent trials (unpublished research, Robson, 1990) prove successful in the long term this will allow the wider application of spray techniques to be used to paint out the ground colour over much larger areas and fast cure under ultraviolet light providing the evaporation of the solvent is ensured before curing.

It is important to match the refractive index of the adhesive to that of the glass to be bonded to effect an invisible repair (Tennent, 1984) The Norland optical adhesives provide a range of refractive indices which can be successfully used to bond glass of different strengths and refractive index ranging from the more substantial to the severely weakened (eg architectural glass). The RI of NOA61 and NOA63 under discussion is 1.56. In addition that of NOA68 is RI 1.54 and the lowest refractive index in their range of optical adhesives is 1.524 (NOA65). In this instance the cured resin is very flexible and is designed to minimise strain. Provided the glass is mechanically sound and a sample is tested for any sign of crizzling or fading of the original hand painting on exposure to ultraviolet light, this adhesive can be suggested for the lightweight bonding of extremely fragile thin sectioned glass. The latter adhesives are currently under investigation and the results will be published when a full assessment of their suitability has been completed.

(Unpublished research, Robson, 1990).

The shelf life of the liquid is at least four months if kept at room temperature in the original container. This can be extended if refrigerated.

The toxological properties of NOA61 and NOA63 have not been thoroughly investigated. At present the manufacturers are not aware of any severe hazards in using this material. As with any organic chemical, care should be taken in handling it. Prolonged skin contact should be avoided and affected areas should be washed well with soap and water. If accidental eye exposure occurs, flush with copious amounts of water and seek medical attention. Use in a well ventilated area.

CONCLUSIONS

The acrylic resins, with their wide ranging and diverse applications have had in the past, a tremendous influence on solving many problems in the conservation of antiquities. For as long as they fulfil all criteria for present day conservation, they will remain in common use. The more recent development of a one part ultraviolet curing acrylic system is but one example of progress in this field of research. Lateral communication between conservators, research chemists, technologists and art historians will ensure that conservation problems and their possible solutions are successfully matched for long term consideration.

The conservator will extend the natural life expectancy of antiquities some of which were never intended to last very long, and yet their natural wastage would entail the loss of important cultural and technological information about the people that crafted and used them.

With the skills of modern scientific research, information technology has progressed from written documentation to computerisation where information is permanently stored and more easily retreived on computerised records.

Long term preservation of our heritage today is thus ensured for the future. Conservators will need to face up to these new technological preservation resources which will, no doubt, pre-empt the use of interventive conservation on some Museum artefacts.

Thus Data Storage will provide a safer passive method for long term preservation of knowledge. With technological advances, the three-dimensional artefact may also be fully documented into a two-dimensional system as a rotating colour representation of the antiquity. This may obviate the need for extensive environmentally controlled storage facilities.

Any archival collections of machine readable databases do, however, require the use of an appropriate machine to read them which will necessitate the updating of software.

The traditional and scientific skills of the artisan/conservator may too, only be required for long term documentation as the electronic auto ego becomes the power tool for recording entire data. Society information thus safely recorded becomes a vision for the future. Realising this goal requires a dedicated cooperative effort of the related custodial, preservation, technological and scientific disciplines.

In the Post Thatcher era, if municipal museums make the decision to computerise totally, information on their antiquities, the collections themselves may need to be privatised for their long term conservation .... After all under museum conditions, the antiquities are rarely handled or even displayed and thus the significant difference to the present situation is one of aesthetics, in being able to visually appreciate actual artefacts. For the discerning scholar decisions will be required for how detailed should be the data stored, how cost effective the preparation and should he have the right of access to view the actual antiquities in private collection...........?

While scientific issues remain central to the conservation profession after six decades since its birth, today, as resource allocation, politics, and public awareness help shape the agenda for conservation, a review of contemporary conditions seems needed to help secure the future of the world's cultural heritage .... if progress in conservation's scientific realm is to have impact, it must be matched by increased public enthusiasm and political support.(Conservation, G.C.I. 1991)

ACKNOWLEDGEMENT

The author wishes to thank the following for their discussion, encouragement and support: David Bailey, Leo Biek, Debbie Boulton, Pat Brown, Andrew Carson, Eileen Collins and Christopher Radbourne, Anne Hudson, Hilde Smith and Pauline Timmins; and to Wendy Firmin, Editorial critic, Bunny Warren, and Kim Sumner Secretariat, Kevin and Robert Brown, Directors, Advance Office Services, UK for practical assistance and professional advice.

REFERENCES

Angst, W. Problems with Objects of Historic Significance, American Institute for Conservation Preprints, 1979

Biek, L. Protective Coatings for silver, Museums Journal, 1952, 60,61.

Beecher, E.R. The Conservation of Textiles, In UNESCO 1968

Horie, C.V. Materials for Conservation, Butterworths. 1987

Organ, R.M. The Consolidation of Fragile Metallic Objects, In Thomson (ed) 1963.

Plenderleith, H.J. The Conservation of Antiquities and Works of Art, Oxford University Press. 1956

Rixon, A.E. The use of new materials as temporary support in the developments and examination of fossils, Museums Journal, 55, 1955, 54-58.

Robson, M.A. "The Use of Aqueous Polymer Emulsions in the conservation of Ethnographic Material: A New Method for Supporting Fragile Woven Fibre Decoration", In Symposium 1986 The Care and Preservation of Ethnological Materials (eds R Barclay, M Gilberg, J.C. McCawley and T. Stone), Canadian Conservation Institute 1986.

Robson, M.A. The use of Norland Optical Adhesives in the Repair and Restoration of Fractured Glass Beads on Ethographic Matierals, ICOM Ethnographic Newsletter, 1987

Robson, M.A. Methods of Restoration and Conservation of Bronze Age Pottery Urns in The British Museum, Early Advances in Conservation, British Museum Occasional Paper No. 65. 1988

Robson, M.A Interior Displays of Restored Stained and Vessel Glass: A Call for suitable Cold Curing Paint Systems for 'Repaints' in situ The Stained Glass Magazine, 1988

Robson, M.A. The Use of Norland Optical Adhesives NOA61 and NOA63 in the Repair and Restoration of Enamelware and Ormolu, CGCG (UKIC) Occasional Paper 1989

Robson, M.A. Historical Applications of the Use of Synthetic Resins for Passive and Interventive Conservation of Fragile Woven Vegetable Fibres and collagenous skins. Atlanta, USA 1990 Publication forthcoming.

Robson, M.A. "One Part Acrylic U.V. Curing Systems with Fillers for Restoration of Porcelain and Enamelware, CGCG (UKIC) West Dean College, 1990

Robson, M.A. The Long Term effect of surface treatments on the properties of cellulosic and ligneous museum artefacts, New Orleans, USA, 1991

Shorer, P.H.T. Soil Section transfers : a method for the transfer of an archaeological soil section on to a flexible rubber backing. Studies in Conservation, 9, 1964 74-77.

Tennent, N.H. and Townsend, J.H. The Significance of the Refractive Index of Adhesives for Glass Repair, In Brommelle et al (eds) 1984.

Terwen, P.A. The mending of Stained Glass; A Technical Instruction, In Tate el al (eds) 1983.

Werner, A.E.A. Plastics aid in Conservation of Old Paintings, British Plastics, 25, 1952, 363-366.

Action of Light on Dyed and Pigmented Polymers

Norman S. Allen

CENTRE FOR ARCHIVAL POLYMERIC MATERIALS, FACULTY OF SCIENCE AND ENGINEERING, MANCHESTER POLYTECHNIC, CHESTER STREET, MANCHESTER M1 5GD, UK

1 INTRODUCTION

Although the first use of colour in the form of dyes and pigments by man is lost in the midst of antiquity evidence is readily available to show that their use operated in the earliest times. Colourants were used to stain not only the human body but also the walls of caves and in the first century A.D., Pliny described a process in use in Egypt which involved the use of what we now know as 'mordanting'[1]. Even in these early times it was known that different colours and hues could be obtained through the use of different metals with a single dye chromophore. Some of the earliest dyes were a luxury such a Murex and Purpura (Tyrian Purple) and yet unfortunately extremely unstable to light.

One of the major problems in this field was the establishment of some method of standardisation for dye and pigment stability especially with regard to textiles. Some of the earliest attempts in this regard were carried out by various Guilds in Scotland and England in the Middle Ages and also in France in the 1300's dyes were classified into low and high fastness grades. However, it was not until 1891 that a British Association set about recording and grading the stability of dyes using a standard. Here dyes were classified into very 'very fugitive', 'fugitive', 'moderately fast', 'fast' and 'very fast'. Thus, all new dyes made at that time had to be graded according to this scale. Today the present standard test method is the ISO Blue Wool method[2] where the test pattern is mounted under glass facing south at an angle of 50-60° to the vertical and is exposed alongside a series of eight blue wool standards. The latter are prepared under standard conditions and are graded in terms of their fastness from one (low fastness) to eight (high fastness) on a geometric scale. The samples and standards are then partly covered and irradiated in sunlight. The samples are compared to that of a special Grey-Scale card consisting of five selectively graded squares of progressive depth. The test pattern is considered to be faded when its loss of colour is judged to be visually equal to that of the initial square and a given member of

the series, usually number three. The pattern is then given
a fading rate which is equal to one of the blue wool
standards which has faded to the same extent. However,
despite the problems and establishment of standards methods
of assessing dye fading the phenomenon itself in both
solution and polymeric media encompasses a number of
problems of technological interest and importance. For
example, dyed textiles may fade or change colour to
commercially unacceptable shades or laser dyes may loose
their efficiency after only a short period of continuous
pumping. On the other hand, in the world of conservation,
both old and new textiles, paintings and prints may fade
and/or degrade to a point where they are no longer of value
in terms of their original quality.

Most of the problems associated with dye fading are, in
fact, due not only to the structure and photochemical
properties of the dye and pigment itself but also the
subsequent interactions that the photoexcited dye molecule
may undergo with its environment. As indicated above some
dyes and pigments, even with a high light stability, may
sensitize the photochemical breakdown of the polymer. In the
textile world this is known as phototendering, and, in many
cases, is relevant to the mechanisms of dye fading.

The above problems may be solved by two approaches. The
development of dye structures that are capable of
dissipating the energy which the dye absorbs from the
radiation in some harmless way while the second involves the
incorporation of appropriate additives into the dyed or
pigmented medium which will either remove the harmful light
energy (absorption) by some effective means or remove and/or
destroy active free radicals and products. Both may be
feasible in the world of conservation depending upon the
type of article requiring protection. Both processes
however, necessitate a detailed understanding of the
analysis and photochemistry of the dyes used. Unfortunately,
many new developments in the dyestuffs field are hindered by
light stability problems and in the conservation world at
far to often a considerable cost.

This article will deal with all aspects of dye and
pigment stability and their influence on the polymer matrix
together with the effects of environmental parameters. With
regard to the latter oxygen and moisture are crucial and is
supported by early work described in 1888 on fading of
artists pigments[3].

2 PHYSICAL FACTORS

Spectral Light Sources

The light fastness properties of dyes is often carried
out under accelerated test conditions utilising Xenon or
high pressure mercury lamps. A number of commercial systems
are available to which the reader is referred for future
reference[4]. In terms of dye fading the nature of the light

source is critical in controlling the rate. The UV content and the heat of the light source can accentuate dye fading and should wherever possible be removed for displays. Some dyes and pigments are also sensitive to visible light in their main absorption band. Elimination of all light is obviously impractical for display purposes but filtered lighting is possible and in this regard for highly sensitive materials yellow/orange lights may be quite suitable without loosing too much effect.

Atmospheric Composition

The effect of atmosphere is a rather mixed and complex subject and often relates more toward the importance of singlet oxygen in dye fading. Numerous articles have dealt with this somewhat controversial subject and will be covered later. However, it is worthwhile noting that while oxygen is found to promote the fading of dyes in solution, studies on polymer films and textiles generally show that oxygen impairs fading due to quenching of the excited state of the dye by the ground-state molecular oxygen. In the latter case it should also be realised that the polymer will also photooxidize and under these conditions will preferentially react with the oxygen on the surface of the material.

Humidity

The importance of humidity in dye photofading is another complex subject which is closely interrelated to the influence of oxygen and the role of singlet excited oxygen. Generally, an increase in humidity will decrease the light stability of a dye but the effect is dependent very much on the nature of the polymer[4-6]. In textiles, absorbed moisture apparently swells the fibres thus enabling air to penetrate the intermolecular pores more readily[4,5]. However, dyes on wool have a lower humidity sensitivity and since this fibre swells more than cotton then some other type of interaction must be involved. In fact, it is now known that in protein structures whilst a photoreductive type mechanism is operative oxidation takes place in other polymeric materials. In this regard dye fading on cotton has been found to be impaired under humid conditions through the use of an after-treatment of potassium thiocyanate and cadmium sulphate[7]. The photohydrolysis of reactive dyes on cotton and nylon 6,6 has also been found to be impaired through the use of the same respective compounds[8]. This effect is demonstrated in Table 1 for the latter study where it is seen that in cotton at 0 and 100% RH the absorbance of dye extracted from the fabric is much less after treatment with potassium thiocyanate while in nylon 6,6 the cadmium sulphate treatment gives a lower absorbance. In the former case hydroxyl radicals are active in the fading and hydrolysis of the dye which are scavenged by the thiocyanate while in the case of nylon polymers photoelectron transfer is important in dye fading and the cadmium sulphate is a well known electron scavenger.

Thus, it is evident that different polymers behave in different ways each requiring a different approach. The photofading of dyes in cellulose and gelatin films illustrated in Figure 1 is a further example of this effect[9]. The results show that while sealing greatly retards oxidative fading of a sulphonated dye in cellulose it has little effect on the reductive fading of a similar dye in a protein substrate.

Table 1   Corrected Absorbancies of Extracts From Fabrics
           Dyed With CI Reactive Blue 161 After Irradiation
           for 45h in a Microscal Unit

| | Cotton | | Nylon 6,6 | |
Reagents	0% RH	10% RH	0% RH	100% RH
None	0.043	0.074	0.444	0.445
KCSN	0.027	0.055	0.435	0.411
$CdSO_4$	0.049	0.107	0.392	0.344

Reproduced from ref. (8) with permission.

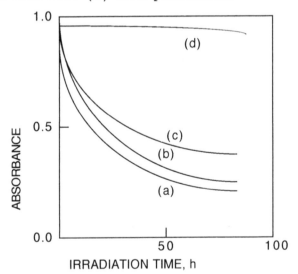

Figure 1   The photofading of CI Acid Red 88 in gelatin
(a) open to atmosphere and (b) sealed and photofading
of CI Direct Blue in cellulose film (c) open to atmosphere
and (d) sealed. Reproduced from ref. (9) with permission.

Temperature

As expected for any chemical reaction an increase in temperature increases the fading rate of a dye. Experiments on this parameter have been utilised mainly to study the effects of oxygen diffusion and aggregation (the ability of the dye/pigment to exist in either single particles or to form conglomerates) on dye light stability. In one study the

activation energies for dye fading on a series of polymer
films were found to follow the order wool < cellulose
acetate < cellulose triacetate < nylon which, in fact, is
the inverse of the order of moisture regain[10]. This suggests
that the ease of diffusion of moisture and/or oxygen are
initial factors. However, both these factors can influence
the state of aggregation of the dye in the respective
polymer and a closer examination of the results indicates
that this may well be the case.

Aggregation and Dye Concentration

Much evidence has accrued over years to indicate that
the aggregation or physical state of the dye in a polymer
substrate is an important parameter in controlling
photofading[4]. Generally, aggregated dyes exhibit a much
higher lightfastness than monodisperse dyes. Dyes in more
amorphous polymers tend to display a higher lightfastness
than when present in a crystalline polymer. This has been
confirmed with experiments where the porosity of regenerated
cellulose was altered[11]. Associated with dye aggregation is
the filter effect which becomes more dominant as the
concentration of the dye in the polymer is increased. In
cellulose dyes appear to build-up in extended mulilayers
rather than as discrete particles. Under these conditions
the inner layers of dye would be protected from the incident
light through attenuation by the outer layers of dye and
local fading will occur only on the surface of the polymer.
However, it may still be argued that dye fading would still
be evident even though it may only be on the surface. This
effect is demonstrated in Figure 2 for three types of
situations. In case (a) the dye distribution is uniform with
dye concentration and there is a constant rate of fading
while in case (b) the particle size is increasing and the
fading rate will decrease and in (c) there is an
unsymmetrical particle growth again decreasing the dye
fading rate in the direction of illumination.

We now know that dyed-polymer systems are much more
complex. Many polymer systems contain a variety of impurity
chromophores produced during manufacture such as
hydroperoxides and carbonylic groups which may interact with
the dye and reduce its lightfastness at low concentrations.
As the dye concentration is increased it will quench the
activity of the chromophores and hence dye stability will
increase. The photofading rate of an acid yellow dye in
nylon 6,6 film is seen to decrease with increasing
concentration in Figure 3.

Dyes fade in various ways and it would in this regard
to illustrate these effects which have been classified into
five types shown in Figure 4. Curve I is classed as first or
second order fading and is associated with dyes in a
molecular dispersion or small aggregates. Curve II has an
initial fast rate followed by a slow zero order rate and is
quite common. It is due to a mixed molecular and particulate
dye dispersion. Curve III shows a constant rate of fade and

is due to an entirely particulate dye. It occurs with pigments and fast dyes. Curve IV has a negative initial fade due to disintegration of the dye particles and occurs quite often with fast dyes. Curve V shows that the dye is fading faster with time due to the continuous breakdown of dye particles. It is characteristic with insoluble dyes in polymers especially cellulose.

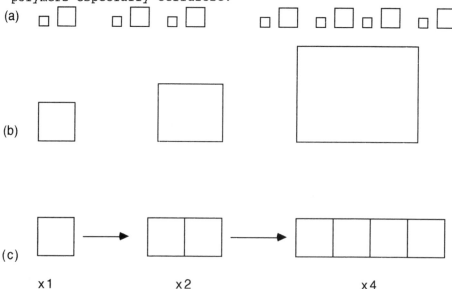

Figure 2 Idealised systems of growth of colourant particles with increasing concentration in the substrate in the ratios 1:2:4.

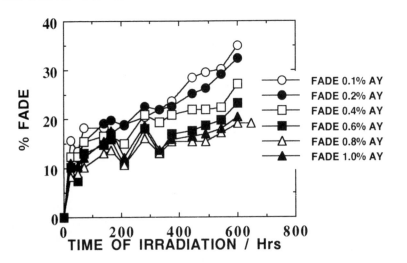

Figure 3 Percentage fade of CI Acid Yellow 135 in nylon 6,6 film at concentrations 0.1 to 1.0% w/w during irradiation in a Microscal Unit.

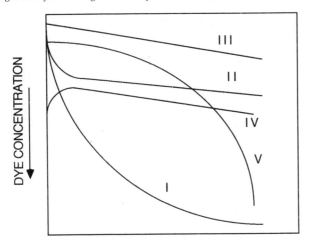

IRRADIATION TIME, h

<u>Figure 4</u> Typical fading rate curves of colourants in
polymers. Reproduced from ref.(4) with permission.

3 MECHANISMS OF DYE FADING AND PHOTOTENDERING

The mechanisms of dye fading are complex processes and often
dependent upon the structure of the dye and the nature of
the polymer. In this article the mechanistic features will
be highlighted from a general point of view together with
some examples of methods of stabilisation. Three main
classes of dye systems will be considered based on the
anthraquinone, azo and triphenylmethane chromophoric
structures 1 to 3 shown below. Associated with the fading
processes of dyes and pigments in polymers is the influence
of the colourants on the stability of the polymer. Here, the
dye or pigment in its photoexcited state will undergo
reactions causing a reduction in the molecular weight of the
polymer (chain scission) and a consequent loss of physical
properties such as tensile strength. Alternatively, some
colourants will actually crosslink the polymer system
inducing brittleness. The mechanisms involved here are
closely inter-related with the mechanisms of dye fading and
will therefore be discussed first.

ANTHRAQUINONE

AZO

TRIPHENYLMETHANE

<u>Structures 1 to 3</u>

Dye and Pigment Phototendering of Polymers

Phototendering occurs when a dye or pigment sensitizes, accelerates or catalyses the breakdown in molecular structure of a polymer substrate or binder material. The term breakdown may involve a reduction in molecular weight or possibly even crosslinking of the polymer. This area has unfortunately, had a history of strong controversy with regard to the mechanisms involved. In recent years, however, interest in this area has declined although there are still many reports of sensitized photodegradation but the mechanistic content is sparse. Over the years two general mechanisms have evolved to account for the dye sensitized photodegradation/oxidation of polymers both of which are relevant to the fading mechanisms of dyes. The first originated from the early work of Egerton[12,13] who proposed a mechanism in which the photoactive excited state of the dye is quenched by ground state molecular oxygen, to produce active singlet-oxygen. The singlet oxygen will then supposedly react with the polymer or water to form hydroperoxide and/or hydrogen peroxide, both of which may then induce further oxidative breakdown of the polymer:

$$D \longrightarrow {}^1D^* \longrightarrow {}^3D^* \tag{1}$$

$$ {}^3D^* + {}^3O_2 \longrightarrow D + {}^1O_2 \tag{2}$$

$$ {}^1O_2 + \text{P-H (Polymer)} \longrightarrow \text{Oxidation Products} \tag{3}$$

$$ {}^1O_2 + 2H_2O \longrightarrow 2H_2O_2 \tag{4}$$

$$ H_2O_2 + \text{P-H} \longrightarrow \text{Oxidation Products} \tag{5}$$

Scheme 1

Reactions (1) to (3) occur primarily under dry conditions while (4) and (5) occur under wet conditions.

The second mechanism was proposed in 1949 by Bamford and Dewar[13,14] and postulates an initial interaction between the photoactive dye and the substrate by a process of hydrogen atom abstraction. This creates a free radical centre in the polymer which is then attacked by oxygen to give an active peroxy radical. The presence of alkali was found to accelerate the process and under these conditions the photoactive dye is believed to abstract an electron from the hydroxyl group to give active hydroxyl radicals.

Reactions (6) to (10) occur under dry conditions while (11) and (12) occur under moist conditions. Later a third mechanism was proposed[17-19] which essentially agrees with the hydrogen atom abstraction theory but instead of the electron transfer step the scheme included the following step to account for the accelerating effect of water:

$$D \longrightarrow D^* \tag{6}$$

$$D^* P\text{-}H \longrightarrow DH\cdot + P\cdot \tag{7}$$

$$P\cdot + O_2 \longrightarrow PO_2\cdot \tag{8}$$

$$PO_2\cdot + P\text{-}H \longrightarrow PO_2H + P\cdot \tag{9}$$

$$PO_2\cdot + D\text{-}H\cdot \longrightarrow PO_2H + D \tag{10}$$

$$D^* + DH^- \longrightarrow D\cdot^- + HO\cdot \tag{11}$$

$$HO\cdot + P\text{-}H \longrightarrow H_2O + P\cdot \tag{12}$$

Scheme 2

$$D^* + H_2O \longrightarrow DH\cdot + HO\cdot \tag{13}$$

as well as the step:

$$DH\cdot + O_2 \longrightarrow D + HO_2\cdot \tag{14}$$

to account for what is known as regeneration of the dye.

Scheme 3

Over the years numerous research papers have appeared in the literature purporting to support either type of mechanism, much of the evidence being based on model system studies which bear no relation to the real polymeric system. To date the relative importance of each type of mechanism has not been resolved and little progress appears to have been achieved with both occupying some degree of importance in the overall process.

Many phototendering dyes are based on the anthraquinone nucleus, in particular vat dyes and these have been the subject of numerous investigations. A typical phototendering dye is CI Vat Yellow 26 of the structure 4 with several examples of weak and strong tendering dyes being shown in Table 2 in viscose and nylon 6 polymer.

Another well-known phototendering dye is 2-piperidinoanthraquinone of the structure 5 which operates by scheme 2 above. In Table 3 it is seen that whilst this dye sensitizes the photochemical breakdown of nylon 6,6 it has no influence on cellulose triacetate. The 1-piperidinoanthraquinone is seen to be a non-tendering dye due to its high stability through the formation of an intra-molecular hydrogen bond between the carbonyl group and the methylene group of the piperidine substituent. This will be discussed further below.

Structure 4

Table 2   Effect of Anthraquinone Vat Dyes on the Sunlight
          Degradation of Fibres

Dye   Loss in Tensile Strength (%)

| | 100% RH | | 0% RH | |
	Viscose	Nylon	Viscose	Nylon
Cibanone Yellow R	80	85	38	73
Cibanone Orange R	52	67	30	65
Caledon Yellow 5G	46	86	40	75
Cibanone Orange 6R	31	20	22	34
Caledon Red BN	11	15	7	15
Caledon Jade Green G	8	8	8	8

Reproduced from ref. (4) with permission.

Structure 5

Table 3   Influence of Piperidinoanthraquinone Dyes on the
          Photodegradation of Nylon 6 and Cellulose
          Triacetate Yarns

| Sample | Loss in Tensile Strength (%) | | Cellulose Triacetate |
| | Nylon | | |
	75 h	150 h	150 h
Undyed	1	9	3
1-PiperidinoAQ	2	10	1
2-PiperidinoAQ	10	23	1

Adapted from ref. (4) with permission.

From all this work the basic facts of dye phototendering
were established and to date they still have not changed.

1. The active tendering dyes are mainly among the yellow and orange dyes, although there are some reds and blues that are active, while there are also some non-tendering yellow and orange ones.

2. The presence of oxygen is essential for significant tendering to occur.

3. The presence of water vapour has a marked accelerating effect on the tendering rate, particularly with cellulosic fabrics.

4. The nature of the fabric is important. Fabrics made from silk, viscose rayon, cotton and nylon are readily degraded, while with wool degradation is virtually non-existent.

5. The degree of aggregation of the dye is important but the overall effect differs with different types of dye and the aftertreatment of the fabric.

6. Fianlly, and this one factor among all others is probably the most important to the dye manufacturer, there is no obvious correlation between chemical structure and tendering activity.

Some early attempts were made to overcome this problem through the use of stabilisers and in this regard certain types of metal complexes were found to be effective. Textile fibres soaked in alcohol/water solutions of these stabilisers retained their tensile properties. Such complexes will be discussed later on dye fading inhibition.

Certain types of pigments such as the cadmium yellows and reds, ultramarine blue and titanium dioxide will also accelerate the photodegradation of polymer media. Here the mechanisms are more difficult to establish due to the inert nature of the pigment. However, much of the work on titanium dioxide pigments has centred on the two crystalline modifications anatase and rutile. For modern applications they are coated with silica/alumina in order to reduce this activity and to improve dispersion properties on polymer processing. Anatase is known to be a powerful phototendering pigment especially in nylon fibres where it is used as a delustrant[13]. Although several variations of mechanisms have been proposed over the years it is generally agreed that many inorganic pigments owe their degree of activity to the presence of surface active hydroxyl radicals or moisture. Singlet oxygen has also been proposed by some workers although the evidence is weak[13]. Scheme 4 shows that reactive hydroxyl radicals present on the pigment particle surface are separated after the absorption of light. During the irradiation an exciton is formed (p) which immediately reacts with hydroxyl groups present on the pigment surface[13]. A hydroxyl radical then separates and an oxygen molecule attaches itself to the residual titanium (IV) ion

and abstracts an electron to produce $O_2\cdot^-$ ions. On reaction
with water, perhydroxyl radicals are produced and the
pigment is restored to its original state while the OH^- and
$HO_2\cdot$ attack the polymer.

Interest in pigment effects on polymers with regard to
durability often centres on paint films where one observes
the phenomenon of chalking on the surface and the loss of
gloss[13].

Dye Fading Mechanisms

Azo Dyes. Azo dyes will either undergo photooxidation
or photoreduction reactions depending upon the nature of the
dye structure, polymer and the atmosphere[4,13]. Substituent

$$TiO_2 \xrightarrow{hv} e' + (p) \tag{15}$$

$$OH^- + (p) \longrightarrow HO\cdot \tag{16}$$

$$Ti^{4+} + e' \longrightarrow Ti^{3+} \tag{17}$$

$$Ti^{3+} + O_2 \longrightarrow [Ti^{4+}O_2^-]_{adsorbed} \tag{18}$$

$$[Ti^{4+}O_2^-]_{adsorbed} + H_2O \longrightarrow Ti^{4+} + HO^- + HO_2\cdot \tag{19}$$

Scheme 4

effects are well-known with azo dyes with electron
withdrawing groups such as CO_2H, NO_2 and CI reducing
lightfastness while electron donating groups such as OH,
CH_3, C_6H_5 and OCH_3 will enhance lightfastness. The former
groups are believed to reduce the strength of the azo link
in photolysis. Scheme 5 compares the general mechanisms of
reduction and oxidation of azo dyes.

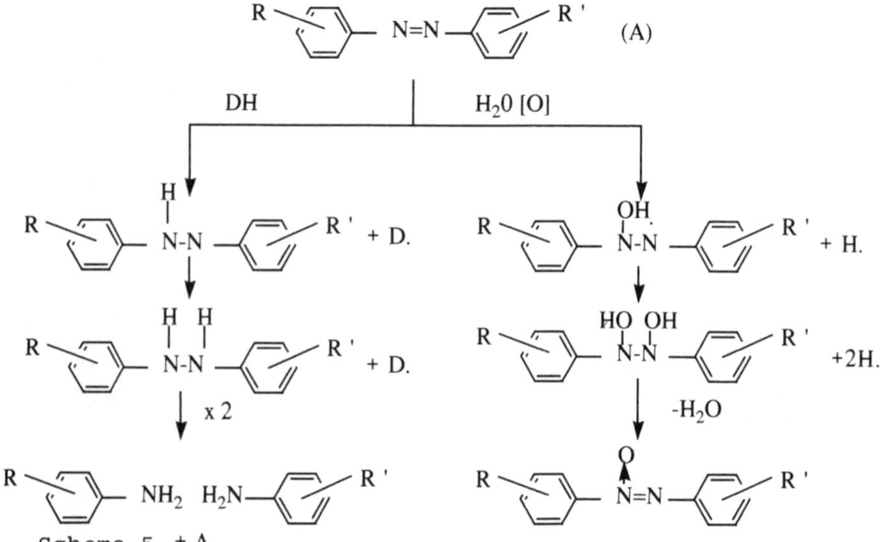

Scheme 5 + A

It should be pointed out that although the reductive reaction route has been confirmed especially in polymers like silk and protein the oxidative route has not been verified experimentally[18]. In the reductive route the azo link abstracts a hydrogen atom from the polymer environment to give a hydrazo compound and eventually a series of anilines as products while the oxidative route produces an azoxy group.

More recently the oxidative route has been shown to involve active singlet oxygen as shown in scheme 6 where the dye in this instance produces naphthaquinone as a product. The fading of azo dyes has been found to be impaired through the use of a singlet oxygen quencher called DABCO (1,4-diazabicyclo-[2,2,2]-octane[18].

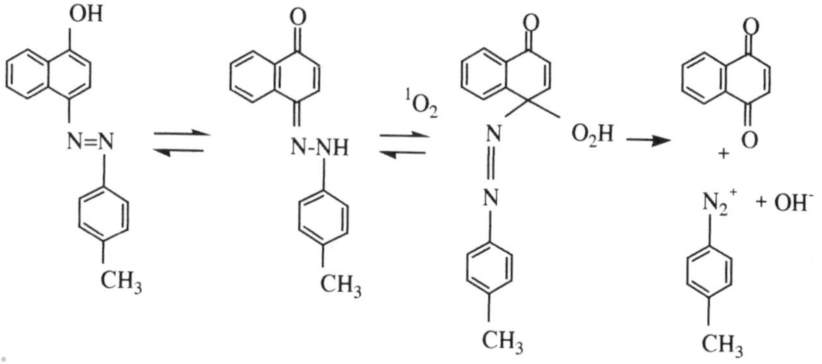

Scheme 6

Anthraquinone Dyes. Commercial dyes based on this chromophore are generally fast to light although they do, in certain cases have the ability to photo-sensitize the breakdown of the polymer as discussed above. Again these dyes will undergo either oxidation or reduction as discussed above in schemes 1 and 2 with singlet oxygen being important in the former case. Much of the mechanistic work has been carried out on alkylamino substituted dyes where dealkylation is the primary step in photooxidation[13]. From a structural point-of-view the position of substitution and nature of the group is important. Thus, substituents in the position α to the carbonyl group on the anthraquinone chromophore that are capable of giving rise to an intra-molecular hydrogen bond such as a hydroxy group shown in scheme 7 are the most stable. Here absorption of light energy converts the dye structure form a stable keto form in the ground-state to a less stable enol form in the excited state. This enol form will then give up its absorbed energy rapidly in the form of heat thus preserving the dye structure intact. The hydrogen bond in the case of a hydroxyl group is strong while that for an amino group is weaker and so the dye will have a lower light stability. Some examples of lightfastness values are shown in Table 4

in polyester and nylon fibres. Thus, all 1-
hydroxyanthraquinone dyes are seen to have a high
lightfastness in polyester (7-8) while all the 1-amino
derivatives are lower (4-6). In nylon the lightfastness
values are even less than in polyester and this is due to
the strong polarity of this type of polymer medium. Here,
the hydrogen bonds in the polymer itself will compete for
those in the dye and so reduce its stability to light.
Furthermore, as discussed previously there is also a
stronger possibility of photoreduction occurring in this
polymer due to electron transfer or hydrogen atom
abstraction reactions. Such processes are not possible in a
polyester environment. All 2-substituted anthraquinone dyes
are not commercial and have very low lightfastness due to
there being no possibility of forming an intra-molecular
hydrogen bond. Dyes with low lightfastness are often highly
fluorescent and give high concentrations of free radicals.
The latter are quite easily observed through the use of a
technique called conventional microsecond flash photolysis.
This monitors changes in the absorbance of a dyed medium
after a short pulse of light on the microsecond time scale
allowing the characterisation of free radicals to be made.

Another structural factor influencing the stability of
anthraquinone dyes is the ability of the dye to dissipate
its absorbed energy from light through the rotation of bulky
groups such a benzene rings (phenyl groups). This is
illustrated in Structure 6 where the dye has a lightfastness
value of 7 in a polyester environment.

KETO-FORM   ENOL-FORM

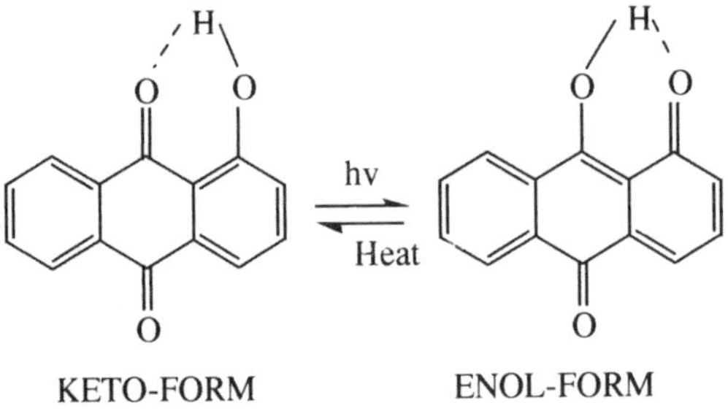

Scheme 7

Table 4   Lightfastness Values (ISO) of Anthraquinone Dyes in
           Polyester and Nylon 6,6 Fibres at a 1/1 Depth

Dye	Lightfastness	
	Polyester	Nylon 6,6
1-Hydroxy	>8	6
1-Amino	4-5	3
1-Amino-2-hydroxy	4-5	3
1-Hydroxy-2-amino	8	5-6

Structure 7

There are many other factors which are important in controlling the fading of anthraquinone dyes. The effect of pH is very important with many dyes being reported to fade 50% faster in low and high pH conditions. Thus, the pH of the fibres especially after washing or treatment will be crucial. Medium pH between 4.5 to 7.5 is desirable to minimise dye fading.

Stabilisation of polymers and solutions containing anthraquinone dyes is feasible through the use of different types of compounds depending upon the mechanistic processes involved. In one such study a blue anthraquinone dye was photo-flashed with xenon lamps ten times (10 kV) in nitrogen saturated 505:50 (v/v) 2-propanol:water solution and was found to fade by 40%. The effect of various stabilisers on this dye (1-amino-4-anilinoanthraquinone-2-sulphonic acid) is shown in Table 5. All exhibit stabilisation with the electron trap, tetracyanoethylene and the nitroxyl radical trap 4-hydroxy-2,2,6,6-tetramethyl-N-oxypiperidine being the most effective agents. A hydroxyl radical trap potassium thiocyanate is seen to be quite effective as are some amines including DABCO. Since oxygen is absent from this medium then the latter agents are likely to operate by quenching the excited states of the dye. The N-oxy radical trap is known to regenerate the dye by reacting with the intermediate semiquinone radical DH· produced in reaction 7 in Scheme 2 above. This is shown in scheme 8 below.

In practice this stabilisation process has been used in the stabilisation of dyed epoxy resins for conservation purposes especially stained glass windows[20] (Figure 5). This data shows that the presence of a light absorber UV 531 has minimal effect on the fading of 1-piperidinoanthraquinone while that of a nickel complex (dibutyldithiocarbamate) is better with the N-oxy radical trap being highly effective and almost completely impairing fading. Here, the nickel complex is known to destroy active hydroperoxides in the epoxy resin which can influence the fading of the dye.

Triphenylmethane Dyes. These are amongst the first of the synthetic dyes to be developed and are still widely used because of their high tinctoral strength (extinction coefficient). Only in recent years has some understanding been gained on the precise nature of elementary primary processes involved in fading. In the presence of oxygen

Table 5   Effect of Various Stabilisers (10^{-3} M) on the %
          Fading of a Blue Anthraquinone Dye in 50:50 (v/v)
          water:2-propanol After 10 Photo-flashes (10 kV).

Stabiliser	% Fading
Control dye	40.0
DABCO	29.3
Ethylenediamine	22.4
Potassium Thiocyanate	23.6
Tetracyanoethylene	4.6
4-Hydroxy-2,2,6,6-tetramethyl -N-oxypiperidine	2.9

Reproduced from ref. (19) with permission.

$$DH\cdot \ + \ \overset{\backslash}{\underset{/}{N}}\text{-}O\cdot \ \text{-->} \ D \ + \ \overset{\backslash}{\underset{/}{N}}\text{-}O\text{-}H$$

Scheme 8

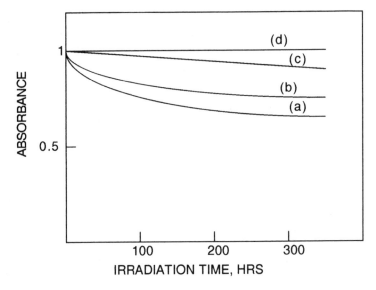

Figure 5   Fading of 1-Piperidinoanthraquinone in a Thermoset
           Epoxy Resin containing (a) no stabiliser, (b) i%
           (w/w) UV 531, (c) 0.1% (w/w) nickel
           dibutyldithiocarbamate and (d) 0.1% (w/w) 4-
           hydroxy-2,2,6,6-tetramethyl-N-oxypiperidine on
           irradiation in a Microscal Unit. Reproduced from
           ref. (20) with permission.

these dyes undergo photooxidation which is initiated by the
ejection of an electron from the photoexcited dye cation
(usually the dye triplet state) to form a radical di-cation
and a hydrated or solvated electron[13,21,22]. The radical di-
cation and electron may recombine to restore the original
dye or they may enter into separate reactions as shown in

scheme 9. Hydrogen peroxide was detected as a product of the fading reaction. These dyes may also undergo a photoreduction process which will involve the formation of a triphenylmethane radical D·.

$$^3D^+ \longrightarrow D\cdot^{++} + e' \text{ (aq)} \tag{20}$$

$$D\cdot^{++} + O_2 \longrightarrow DO_2\cdot^{++} \longrightarrow \text{Product} \tag{21}$$

$$D + e' \longrightarrow D\cdot \tag{22}$$

$$2D\cdot + H_2O \longrightarrow DH\cdot + D^+ + HO\cdot \tag{23}$$

$$e'(aq) + O_2 \longrightarrow O_2\cdot^- \tag{24}$$

$$O_2\cdot^- + H_2O \longleftrightarrow HO_2\cdot + HO^- \tag{25}$$

$$2HO_2\cdot \longrightarrow H_2O_2 + O_2 \tag{26}$$

$$HO_2\cdot + e'(aq) \longrightarrow HO_2\cdot^- \tag{27}$$

$$HO_2\cdot^- + H_2O \longleftrightarrow H_2O_2 + OH^- \tag{28}$$

Scheme 9

As with other classes of dyes the role of singlet oxygen is considered to be important but again the experimental evidence is based on solution studies which bear little or no relation to actual practical polymer situations[23]. The light stability of these types of dyes is markedly influenced by the nature of the polymer. In cellulosics they have extremely poor lightfastness (1) compared for example to acrylics (Orlon) (3)[21]. Such a marked differential in lightfastness is also extrapolated to solution media as shown in Figure 6[21]. Here it is seen that after one single photo-flash in 2-propanol Brilliant Green YN fades completely whereas in acetonitrile there is little change. These data are related in a sense to the lightfastness of the dye in cotton and acrylonitrile respectively which are 1 and 3 on the ISO scale. Radical formation due to the triphenylmethane radical on microsecond flash photolysis is also higher in the former solvent. There are two explanations for this observed effect. The first is that in an acrylic environment the cyano groups are powerful electron traps and will trap the electron in scheme 9 above. The second is that in an acrylic environment radical recombination is encouraged and the dye will be regenerated compared with that in cotton.

In the photofading of triphenylmethane dyes in scheme 9 above the formation of the solvated electron is the key step and should this be scavenged then stabilisation may be achieved. This has been shown to be the case as illustrated in Figure 7. Here the photofading of Malachite Green in polyvinyl alcohol is seen to be inhibited by the presence of a powerful electron trap tetracyanoethylene but, in this case, not the radical trap, 4-hydroxy-2,2,6,6-tetramethyl-N-oxypiperidine[22]. Typical antioxidants such as hindered phenols were also found to be ineffective in this work.

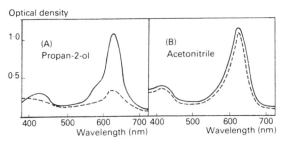

<u>Figure 6</u>   Absorption spectra of solutions of Brilliant Green YN (10^{-6} M) in (A) 2-Propanol and (B) Acetonitrile before (_____) and after (-----) one photo-flash. Reproduced from ref. (21) with permission.

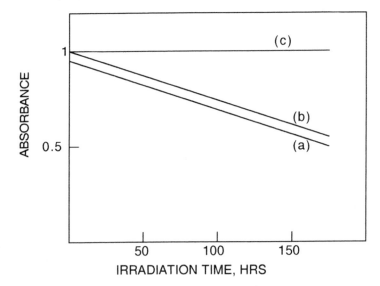

<u>Figure 7</u>   Change in the absorption spectrum of Malachite Green in Poly(vinyl alcohol) film (25 μm thick) containing (a) no additives, (b) 0.1% (w/w) 4-hydroxy-2,2,6,6-tetramethyl-N-oxypiperidine and (c) 0.1% (w/w) tetracyanoethylene during irradiation in a Microscal Unit. Reproduced from ref. (22) with permission.

The formation of the triphenylmethane radical was also found to be inhibited on laser flash photolysis using other electron traps such as nitrous oxide gas and acetone[22].

Remaining with the triphenylmethane dyes another important area is printing where such dyes are used in their leuco forms. Leuco crystal violet is an example of such a system which is utilised in the form of starch cells embedded in paper. The cells contain a Lewis Acid which is released on printing and will regenerate the dye colour[24] as shown in scheme 10. Such a colour is extremely unstable especially in the presence of an acidic environment.

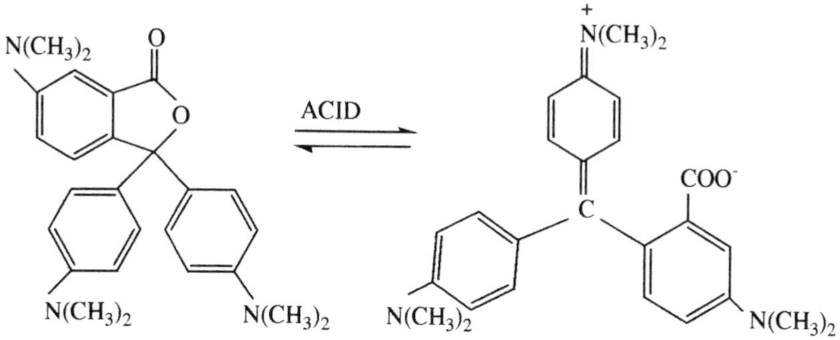

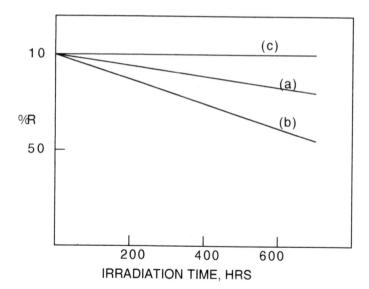

Scheme 10

Figure 8 Percentage Reflectance Spectra Versus Irradiation Time (h) in a Xenotest-150 unit for Paper Prints of Crystal Violet Lactone containing (a) no additives, (b) 0.1% (w/w) 4-hydroxy-2,2,6,6-tetramethyl-N-oxypiperidine and (c) 0.1% (w/w) tetracyanoethylene. Reproduced from ref. (24) with permission.

On fading this type of chromophore in solution
hydroperoxides were found to be important intermediates
and in this regard the presence of a nickel complex
(dibutyldithiocarbamate) was found to be effective in
inhibiting dye fading in an acid medium[24]. The presence of a
radical trap was found to be ineffective. These results are
shown in Figure 8 for actual paper prints using this dye.
This agrees with other model system work[23] for a wide
variety of dyes showing that a practical solution to a
commercial problem is certainly feasible.

4 CONCLUSIONS

Dye fading is a complex phenomenon involving reactions not
only associated with that of the dye but also those
involving the polymer as well as dye-polymer interactions
which are further complicated by the nature of the
environmental conditions. Stabilisation is feasible for many
systems where the aesthetic properties of the system may not
be affected through an appropriate treatment. Failing this
the nature of the environmental conditions for storage need
to be controlled for maximum life of the article.

REFERENCES

1. R. Meldola, *J. Soc. Dyers and Colourists*, 1910, *26*, 109.
2. British Standard 1006, (1978), British Standards
    Institution, London.
3. W.J. Russel and W.de W. Abney, "Action of Light on Water
    Colours","Report to the Science and Art Department of
    the Committee of the Council on Education", 1888, *C5453*,
    HMSO, (London).
4. C.H. Giles and S.D. Forrestor, in "Photochemistry of
    Dyed and Pigmented Polymers", (Eds. N.S. Allen and J.F.
    McKellar) Elsevier Science Publishers, London, 1980,
    Chpt.2, p. 51.
5. C.H. Giles and McKay, *Text. Res. J.*, 1963, *33*, 527.
6. C.H. Giles, G.D. Shah, D.P. Joha and R.S. Sinclair,
    J. Soc. Dyers and Col., 1972, *88*, 433.
7. A. Datyner, C.H. Nicholls and M.T. Pailthorpe, *J. Soc.
    Dyers and Col.*, 1977, *93*, 213.
8. N.S. Allen, K.O. Fatinikun, A.K. Davies, B.J. Parsons
    and G.O. Phillips, *Polym. Photochem.*, 1981, *1*, 275.
9. C.H. Giles, R. Haslam and D.G. Duff, *Text. Res. J.*,
    1976, *46*, 51.
10. C.H. Giles, B.J. Hojiwala, C.D. Shah and R.S. Sinclair,
    J. Soc. Dyers and Col., 1974, *90*, 45.
11. C.H. Giles and Haslam, *Text. Res. J.*, 1978, *48*, 490.
12. G.S. Egerton, *J. Soc. Dyers and Col.*, 1947, *63*, 161.
13. N.S. Allen, *Revs. Prog. Col.*, 1987, *17*, 61.
14. C.H. Bamford and J.S. Dewar, *Nature*, 1949, *163*, 214.
15. J.J. Moran and H.I. Stonehill, *J. Chem. Soc.*, 1957, 765.
16. J.J. Moran and H.I. Stonehill, *J. Chem. Soc.*, 1957, 779.
17. J.J. Moran and H.I. Stonehill, *J. Chem. Soc.*, 1957, 788.
18. J. Griffiths, in "Developments in Polymer

Photochemistry", (Edited by N.S. Allen), Elsevier
Applied Science Publishers Ltd., London, 1980, Vol. 1,
Chapter 6, p. 145.

19. N.S. Allen, K.O. Fatinikun, A.K. Davies, B.J. Parsons
and G.O. Phillips, Dyes and Pigments, 1981, 2, 219.
20. N.S. Allen, J.P. Binkley, B.J. Parsons, G.O. Phillips
and N.H. Tennent, Dyes and Pigments, 1983, 4, 11.
21. N.S. Allen, J.F. McKellar and B. Mohajerani, Dyes and
Pigments, 1980, 1, 49.
22. N.S. Allen, B. Mohajerani and J.T. Richards, Dyes and
Pigments, 1981, 2, 31.
23. N. Kuramoto and T. Kitao, Dyes and Pigments, 1981, 3,
49.
24. N.S. Allen, N. Hughes and P. Mahon, J. Photochem., 1987,
37, 379.

Subject Index